The History of The
GREEN HOWARDS

THE HISTORY OF THE
GREEN HOWARDS

THREE HUNDRED YEARS OF SERVICE

Geoffrey Powell
and John Powell

FOREWORD BY HIS MAJESTY HARALD V
KING OF NORWAY

'Not chance nor accident of mustering makes the
troop, but family and friendship; and this is a very
powerful incitement to valour.' (Publius Cornelius
Tacitus, circa AD 55-120)

MILITARY

First published in Great Britain 1992
by Arms and Armour Press

Reissued in this revised and enlarged edition 2002 by
LEO COOPER

Reprinted in 2015 by
PEN & SWORD MILITARY
an imprint of
Pen & Sword Books
47 Church Street,
Barnsley, S. Yorks, S70 2AS

HB ISBN 978 1 47385 796 4
TPB ISBN 978 1 47385 797 1

Printed and bound in England by
CPI Group (UK) Ltd, Croydon, CR0 4YY

Pen & Sword Books Ltd incorporates the imprints of Aviation, Atlas,
Family History, Fiction, Maritime, Military, Discovery, Politics, History,
Archaeology, Select, Wharncliffe Local History, Wharncliffe True Crime,
Military Classics, Wharncliffe Transport, Leo Cooper, The Praetorian
Press, Remember When, Seaforth Publishing and Frontline Publishing.

For a complete list of Pen & Sword titles please contact
PEN & SWORD BOOKS LIMITED
47 Church Street, Barnsley, South Yorkshire, S70 2AS, England
E-mail: enquiries@pen-and-sword.co.uk
Website: www.pen-and-sword.co.uk

CONTENTS

Above: Shoulder-belt plate of the 19th Regiment worn by Lieutenant J. H. Kirke during the Crimean War.

Above: The Regiment's badge today. It consists of the cypher of HRH Alexandra, Princess of Wales, interlaced with the Dannebrog, inscribed with the date '1875', the Roman numerals 'XIX' below and the whole surmounted by the Coronet of a Princess.

LIST OF MAPS

I was greatly honoured when Her Majesty
Queen Elizabeth The Second asked me to
become Colonel-in-Chief of the Green
Howards. I become the fourth generation of
my family to hold the post, following in
the footsteps of my father, my grandfather
and my great-grandmother. The Green Howards
are indeed a family regiment.

The appointment marks also that close
friendship between Norway and Great
Britain, especially with those counties of
North-East England in which lie the homes
of so many of our soldiers, a friendship
that has existed for many centuries past
but one so greatly strengthened in recent
years.

This book describes the sacrifices made by
the Green Howards in the service of their
country during the past three hundred
years. It is a most interesting story which
highlights the great strength of the
British Army Regimental system.

Harald R

PREFACE

BY

FIELD MARSHAL THE LORD INGE, KG, GCB, DL

Colonel, The Green Howards 1982-1994

First Edition

This Regiment has been well served by its historians. Colonel Geoffrey Powell's predecessors have ably told the story of its early years, of the part it played in the Second Boer War, the two World Wars of this century and the Malayan Emergency of the 1950s. His own previous excellent short volume in that fine 'Famous Regiments' series of regimental histories has also made its contribution.

Until now, however, we have lacked a history of our Regiment available to all in a single volume. Nor have either the inter-war years from 1919 to 1939 or the decades that followed the Second World War been properly researched or described. This is the reason why I asked the author, an old friend and mentor, to write this book.

No one volume history can be comprehensive. To tell the story of our Regiment in all its detail would need a shelf of books. This book then is a distillation of the essence of our Regiment's history, told in a readable manner. Researchers and others investigating a subject or incident in greater detail will have to go back to those earlier books, to the relevant issue of our *Green Howards' Gazette*, or to other sources. To all of these the author provides the necessary pointers.

After reading this book and glancing through its sources, I have been especially impressed by the importance of the *Green Howards' Gazette* as the repository for the details of our history. Without its articles on specific subjects and its routine accounts of unit activities, it would have been all but impossible for this history to have been written. I cannot emphasise too strongly the need for my successors to ensure the *Gazette's* continued publication.

History can be read for both profit and pleasure. The weapons of war continually change, as do the tactics for their use, but the human element changes little. There is much to be learned from these pages, written as they are by a soldier. And it will be a dull individual who fails to be moved to both pride and sadness by this stirring account of the deeds of his predecessors. Above all, of course, this book is a statement of the debt owed by both the nation and the Army to our regimental system.

Second Edition

The great success of the first edition of Colonel Geoffrey Powell's Regimental history has made it necessary to reprint it in order to meet the demand. This has given the opportunity to add a new chapter covering the last ten years. This decade has been a very busy one in which the Regiment continues to play a full part in the nation's history.

It is particularly appropriate that Brigadier John Powell, with his intimate knowledge of the last thirty-five years of the Regiment's life, has joined his father in updating this excellent Regimental history; it is a fine example of the importance of the family in our Regiment's history.

AUTHOR'S INTRODUCTION

The Green Howards with fitting pomp and great zest celebrated their Tercentenary in July 1989, one year late because active operations had prevented their so doing at the right time. By a coincidence, the same had happened half a century before when one of the then two Regular Battalions of the Regiment was serving in a troubled Palestine and the other upon the North-West Frontier of India. Not until July 1939 was the 1st Battalion able to mark that two hundred and fiftieth anniversary in a proper manner; a few more weeks and the outbreak of the Second World War would have caused a further postponement.

That Tercentenary of the founding of the Regiment suggested that an up-to-date history was due, one covering the full three hundred uninterrupted years of service to its monarchs. So it was that in the spring of 1989 I was delighted to be asked by the Colonel of the Regiment, the then General Sir Peter Inge, to start work upon such a book.

Much had already been written about the Regiment either in long out-of-print books or in the near 100 volumes of the Regiment's journal, *The Green Howards Gazette*, published monthly for most of its century of existence but latterly quarterly. Without the work of previous authors, and of the countless thousand contributors (more often than not pressed men) of either articles or unit notes to the *GHG*, the book could not have been written. Especial tribute must be paid to Major M.L. Ferrar, the Regiment's most distinguished historian, founder of both the *Gazette* and the Regimental Museum, as it must also be to Colonel H.C. Wylly, Captain W.A.T. Synge and Brigadier J.B. Oldfield. Details of their books and of others written about the Regiment will be found in the bibliography.

Because I had written an abbreviated history of the Regiment almost thirty-five years ago for the well known *Famous Regiments* series, I had a frame upon which to start work. I made no apologies for using again some of the material from that previous book. Often I had found it difficult to improve upon my previous words; others may feel that they could have done.

In putting together the material for that earlier book, I expressed my deep gratitude to the many Green Howards who had searched their memories and read my several drafts, but I mentioned by name only the late Brigadier T.F.J. Collins and the late Colonel J.M. Forbes, the then Regimental Secretary, two of my principal guides and helpers. The individual who had, in fact, given me even more assistance was the late Colonel A.C.T. White, VC, MC. That fine scholar, with his wide knowledge of the Regiment and its history, was unstintingly helpful. Deep was his affection for the Green Howards from which he had been obliged to transfer to a corps because, as a young married officer in the early 1920s, he could not afford to continue to live in it on his pay. His understanding of human nature in general and that of soldiers in particular was profound. Another of his outstanding qualities was his feel for the English language. In a manner kindly yet critical, Archie directed the footsteps of a tyro author who was tackling with some trepidation his first book. That very modest man insisted that his name should not be mentioned; I felt sure that he would not have minded my doing so after his death. The late Brigadier E.C. Cooke-Collis was another who at the time was generous with his help.

A book such as this is inevitably a co-operative undertaking, one that involves a number of people. Two in particular I thanked. Lieutenant Colonel N.D. McIntosh, the Regimental Secretary, was responsible for the illustrations and several of the appendices; he also provided me with a steady stream of information and sound advice, and he read and commented incisively upon the draft. Secondly, my son John Powell, from the depths of his recent knowledge of the Regiment, wrote the first draft of chapter 12 and produced many of the thoughts for the Epilogue. However, when the Regiment decided that a new edition of the book should be prepared, one in which its story would be brought up-to-date, we agreed that we should do so as joint authors, and that, in so doing, he would be primarily responsible for preparing a new chapter 13 and bringing the Epilogue up to date. The final responsibility was, however, mine.

Others who have commented upon the draft of the earlier edition either in part or in whole were the now Field Marshal Lord Inge, Major P.J. Howell, the late Lieutenant Colonel D.G. St J. Radcliffe, the late Lieutenant Colonel E.D. Sleight, the late Lieutenant Colonel D.M. Stow, and Colonel H.A. Styles. I am grateful to them all, as I was to two then subalterns, Richard Inman and Jason Wright, who delved cheerfully into various archives in the North of England. The other members of the staff of Regimental Headquarters, especially Mr John Goat, were unfailingly helpful; the last-named, in particular, I caused much extra work, all of it tackled in what was his usual cheerful manner. My grateful thanks were also due to Mrs Trish Cavell of the 1st Battalion staff, Major K. Gardner, the late Major W.H.G. Kingston, Mr O. Neighbour and the late Brigadier J.B. Oldfield, all of whom helped me in a variety of different ways. As she had done with all my books, my wife Felicity helped me in innumerable ways reading every page (sometimes twice) and commenting incisively.

The present Colonel of the Regiment, Major General Richard Dannatt, has given his unstinting support to this second edition, and my son and I are especially grateful to him and to Major J.R. Chapman at Regimental Headquarters, Major JF Panton and Mrs Sue Woodall, my son's secretary, who have helped so much in its preparation. In addition, Field Marshal Lord Inge and the commanding officers of the last decade have commented helpfully on the draft of chapter 13.

Tribute was also paid to another old friend, the late Brigadier J.B. Scott, who had the imagination to set up the Regiment's Oral Archives before his death. As a source for the Regiment's more recent history, these proved most valuable; they will be even more so as the years pass. Lieutenant Colonel J.R. Neighbour's transcription of parts of the Erle Papers in Churchill College, Oxford, had added much to our knowledge of the Regiment's early days and saved me much arduous work in those archives; he also gave me useful advice on other matters.

Thanks are also due from both of us to the executors of the late Sir Herbert Read and to Faber and Faber for their permission to use extracts from Sir Herbert's *Collected Poems* and *The Contrary Experience*. We are also grateful to the organizations and artists whose names appear in parentheses after the captions to the illustrations for permission to reproduce their illustrations.

As ever in a book such as this, one that touches almost every continent and spans three centuries, the nomenclature and spelling of places proved a near insoluble problem. We possibly took the easy way out; in most cases adopting the usage at the time the place was first mentioned.

As is their way, librarians and archivists have been unfailingly helpful. Among them were those of the Chipping Campden Branch of the Gloucestershire County Library; the Imperial War Museum; the Liddle Collection of the Edward Boyle Library of Leeds University; the London Library; the National Army Museum; and the Library of the Royal United Services Institute. Major A.E.F. Waldron, the Regimental Secretary of the Middlesex Regiment, Brigadier J.K. Chater, the Regimental Secretary of the Warwickshire Area of The Royal Regiment of Fusiliers, and the staff of the Royal Hospital, Chelsea, also willingly replied to queries.

The Regiment is extremely grateful to its Allied Regiment, the Queen's York Rangers of Canada, for its generous contribution towards the publication costs of the first edition of this book.

In first commissioning this book, Peter Inge gave me the free hand I sought. At no stage have we been subjected to any type of censorship. As I wrote in the preface to my earlier book on the Regiment, 'Memories are easily warped, particularly memories of battles, and in the specialist field of regimental history a further trap awaits the writer: a regiment could never transgress, and consciously or unconsciously, his forerunners often wrote in this fashion'. This suggestion of infallibility is manifestly absurd. As with all regiments, the Green Howards won glory in plenty but at times failed in what they had set out to do. Our aim has been to write an unbiased story; to do otherwise would have shown contempt for the discerning reader. For Green Howards such as ourselves, it has not always been easy to discard the inevitable prejudice in their Regiment's favour, the fruit of near lifelong affection for it. Perhaps we failed to do so. On the other hand there may be those who object to our having told the truth as we saw it, even when it has led to the revelation of the unpalatable and the suppression of doubtful but pleasing legend. We can only say that we tried to write history.

GEOFFREY POWELL

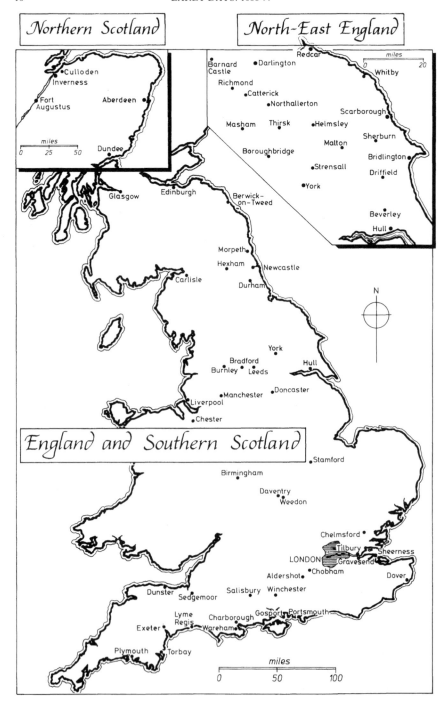

Northern Scotland

Culloden
Inverness
Fort Augustus
Aberdeen
Dundee

miles
0   25   50

North-East England

miles
0   20

Redcar
Barnard Castle
Darlington
Whitby
Richmond
Catterick
Northallerton
Scarborough
Masham
Thirsk
Helmsley
Sherburn
Malton
Boroughbridge
Bridlington
Strensall
Driffield
York
Beverley
Hull

Glasgow
Edinburgh
Berwick-on-Tweed
Morpeth
Hexham
Newcastle
Carlisle
Durham

N

England and Southern Scotland

York
Bradford
Burnley
Leeds
Hull
Manchester
Doncaster
Liverpool
Chester

Stamford

Birmingham

Daventry
Weedon

Chelmsford
Tilbury
Sheerness
LONDON
Gravesend
Aldershot
Chobham
Dover
Dunster
Salisbury
Winchester
Sedgemoor
Lyme Regis
Charborough
Gosport
Portsmouth
Exeter
Wareham
Plymouth
Torbay

miles
0   50   100

1. EARLY DAYS: 1688-97

During the years that followed the Civil War, military rule by Cromwell's majors-general had given the English people a profound distaste for professional soldiers. With the restoration of King Charles II to his throne in 1660, both Houses of Parliament were resolved that any standing army should be limited to a few regiments of Guards for the King's protection, together with the troops needed to garrison coastal and other forts.

When King Charles died, his younger brother, James II, an impetuous and ardent Roman Catholic, succeeded him. Ignoring the prejudices of his fellow countrymen, by 1688 the King had raised a 30,000-strong regular army, housed in a great camp at Hounslow Heath where it overawed London. Officered by too high a proportion of the King's co-religionists, Celtic-speaking Irish peasants were imported to fill badly recruited regiments. Because of such autocratic and ill-considered measures, and fearing the threat to liberty and the nation's established religion, a number of great territorial magnates invited William of Orange, the Stadtholder of Holland, whose wife Mary was King James's daughter, to supplant the unpopular King James.

When Prince William landed at Torbay on 5 November 1688 with 14,000 men, including six British regiments in the Dutch service, there was no resistance. James's unpopular army, among whose leaders was Major-General John Churchill, the future Duke of Marlborough, deserted its master without putting up any resistance. What was to be known as 'The Glorious Revolution' was quickly over.

Fresh and loyal troops were, however, needed to sustain the Protestant cause, and so, a little over three hundred years ago, there came into being Colonel Luttrell's Regiment of Foot, later to be known as The Green Howards.

Francis Luttrell of Dunster Castle, the head of one of Somerset's most ancient and wealthy families, was one of the first of the great West Country landowners to join William's standard. Commissioned by the Prince to raise a regiment of foot, Luttrell did so within three days. On 19 November 1688, as a contemporary writer related, 'The Prince of Orange left the citty of Exeter with his army, and left Mr. Seymor, Governour, with Colonel Luttrell's Regiment to secure it.'[1] In due course it was to take precedence as the 19th Regiment of Foot, the first to be raised in England after Prince William's landing.

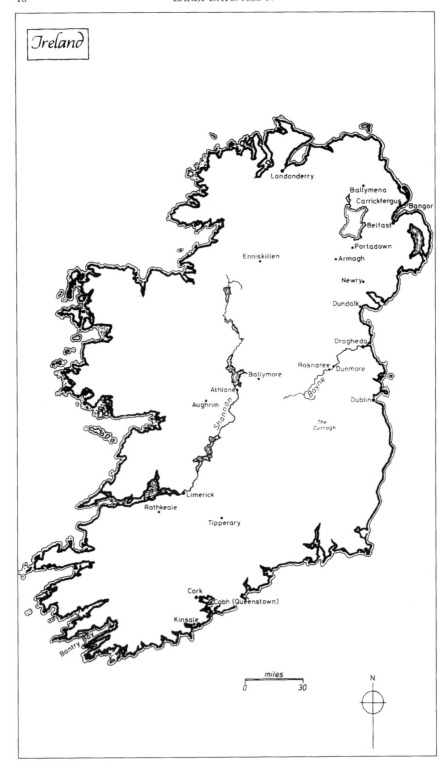

When King James fled the country to take refuge in France, Parliament directed that William and his wife Mary should reign jointly in his place. A petition addressed to the new King the following year reads:

> To the King's Most Excellent Majestie. The humble petition of Coll. Francis Luttrell in behalfe of himself and the rest of the Officers in his Regiment. Sheweth That at your Majestie's happy Arrival in this Kingdom you were gratiously pleased to give a Commission to your Petitioner for raising a Regiment, which He compleated in three days time, and kept the same fourteen days on his owne expence; And that it might be ready to march where your Majestie's service required, your Petitioner clothed the said Regiment at Exeter, which cost neare £1500; in which he was so extremely abused (the Clothes being already wore out) that it is absolutely necessary for your Majestie's Service to cloathe anew. Your Petitioner therefore most humbly Prayeth your Majestie That Whereas there still remains due to the said Clothiers a considerable sume and that its necessary for the Regiments imediate Clothing to direct the Paymaster General, to cleer the said Regiment from the beginning of January last to the first of July, which will be of great use to your Petitioner towards paying for the Old Cloaths, and what remains then of that sume He is willing to pay out of his own Estate, so that the Two pences from that time may goes towards the new Clothing with which and his own obligacon (which he is willing to give) He hopes to have the Regiment in a short time clothed, and fitt to march where your Majestie's Occasions may require.[2]

It is a coincidence that of the few surviving documents relating to the Regiment's early days, two of them discuss uniform. In the Luttrell archives there reposes a tailor's bill of January 1689 asking for payment for 'the lineing of your imbroydered coat, being of richer sattin and much better than the lineing of the other officers £1=6=0d. To pay for blew cloth for your coat, being much better than the other officers, 10/-'.[3] This bill is of special interest. It shows that when the Regiment was raised it was dressed, not in the scarlet introduced in Cromwell's army, but in blue, probably in flattering imitation of King William's famous Dutch 'Blue Guards'. The facings were yellow, the colour of the Luttrell family livery.

It could be claimed that the Regiment's ancestry can be traced even further back. In 1685, the Duke of Monmouth, illegitimate son of King Charles II, had landed in the West Country to raise an earlier flag of Protestant revolt against King James, an expedition that was to end with the slaughter of the Battle of Sedgemoor, Monmouth's death on the scaffold and the ruthless repression of Judge Jeffreys' 'Bloody Azzize' in which hundreds of smallholders, tradesmen and weavers were executed or transported to the sugar plantations of the West Indies.

Francis Luttrell, who had become colonel of his local Militia regiment five years earlier, was deeply involved, but this time on the side of the legitimate government. The Militia (about which more will be said later) had been, since Saxon days and in a variety of different shapes, the main force upon which the monarch could rely for defence against external attack and to suppress insurrection; raised upon the principle that all men were under the obligation to take part in their country's defence, service was part time and the costs met by local levies of various types.

In the first proper engagement of the rebellion Monmouth's men, advancing northwards from Lyme Regis, where their leader had landed, met the Militia regiments of Somerset and Cornwall. Luttrell, confident indeed, reported that the rebels were disordered. The outcome was unhappy. Luttrell's men, together with the rest of the force, broke and fled as soon as the rebels appeared, some discarding their uniforms and weapons in their haste to get away. Not only were they virtually untrained, but they had little stomach for fighting their neighbours and co-religionists. Here and in later engagements the Militia proved useless.[4] The whole of the then small regular army had to be concentrated to crush the rebellion.

There is an apocryphal story that Luttrell and his Militiamen, deserting to Monmouth, marched to join him but arrived just as he was being beaten at Sedgemoor. Seeing how matters were going, they straight away made for home where they turned out their horses to pasture for fear that their wet appearance would betray them.[5]

When Luttrell raised his regular regiment three years later, he could hardly have done so in three days unless his Militia had formed its backbone – men encouraged to enlist in it partly by their loyalty and obligations to the Luttrell family, but also by their strong Protestant sympathies and their bitter memories of the aftermath of Monmouth's rising. Three men, probably neighbours, had been hanged at Dunster after the Bloody Assize.[6]

★★★

Somehow Luttrell's Regiment was reclothed during that summer of 1689 and it moved, first to Portsmouth, then to the Isle of Wight and later into the Citadel at Plymouth. In the meantime, however, King William, threatened by France, was raising further regiments, fourteen in the early months of the year. Most were to be disbanded, but among them was one commanded by Colonel Thomas Erle of Charborough, a Deputy Lieutenant of Dorset and Member of Parliament for Wareham. Erle's Regiment was to be the second root from which grew the Green Howards.

Like Luttrell, Erle had commanded a regiment of Militia on King James's side at the time of the Monmouth rising and, like him, he later transferred his support to Prince William. This is shown by an inscription over an ice-house in the grounds of Charborough Park, built by one of Erle's descendants, which reads, 'Under this roof, in the year 1686, a set of patriotic gentlemen of this neighbourhood concerted the great plan of the glorious revolution with the immortal King William.'[7] As with Luttrell's Regiment, the origin of Erle's can be traced back to this Militia, for at a meeting held on 5 December 1688, 'the Deputy Lieutenants and other Gentlemen' of Dorset resolved to raise two regiments of Militia 'for the Assisting of the Prince of Orange in [this] Glorious Undertaking . . . to be sev'ally Commanded by Colonel Strangeways and Colonel Erle . . .'.[8] Preserved among the documents concerned is one epitomizing a commonplace and very human problem, a letter complaining that 'one Thomas Sturmer of Swanage a Stone Cutter that hath listed himself in yr Company, his Mother being an antient woman, is very much troubled at her sons being goe for a soldier from her and therefore doth make her humble request that you will please to release her son, else "tis said it will break the old woman's heart".[9]

King James's flight must have temporarily removed the Prince of Orange's need for armed support, for on Christmas Eve Erle received orders to 'dismiss the Regiment of souldiers under your Command to their Respective Dwellings (letting them take their Coats, Hats and Stockings) until you shall have further orders from His Highness . . .'.[10] It can be assumed that most of these Militia 'souldiers' were embodied in Erle's regular regiment, raised on 8 March 1689 and soon to be in action.

★★★

By 1688 King Louis XIV, the ruler of France for the past twenty-seven years, in a series of aggressive wars had acquired for his country enormously valuable territories. With varying success, Prince William had defended the Netherlands against these French depredations and, in so doing, had become the focus of Protestant resistance on the continent of Europe. William's seizure of the throne of England was the spark that led to a further outbreak of hostilities in which Britain and France were to fight each other, with short breaks, for the next century and a quarter.

King Louis began by supporting King James in his invasion of Ireland, a country in which his co-religionists formed three-quarters of the population. Landing at Cork on 14 March 1689 and bringing with him some hundreds of officers, French arms and French money, James raised levies and was joined by the Irish Catholic regular regiments stationed in the country. The Protestant Irish, facing disaster and destruction, took refuge in Londonderry and Enniskillen where they resisted James's forces with the courage of desperation in epics which have become a part of the folklore of the Six Counties.

Erle's Regiment was part of the 17,000-strong relief expedition made ready by King William during the summer. On 13 August this force began to land at Bangor, County Down, under the command of the veteran Dutch Marshal Schomberg. After a dozen regiments had been straight away detached to besiege Carrickfergus, which surrenderd on 21 August, the army marched through a devastated countryside by way of Newry to reach Dunkald.

It took no more than a week for Schomberg to discover that he could rely only on his few Dutch and Huguenot regiments. As the Reverend George Story, a chaplain with the expedition, recorded, the new English regiments had been 'mustered and disciplin'd as well as time would allow'. One in four had never even fired their muskets and those that had 'thought that they had done a feat if the Gun fired, never minding what they shot at'. They were no more than mobs of undisciplined boys, commanded by ignorant, negligent and usually dishonest officers. Their weapons were useless, and arrangements for feeding, clothing and pay all but non-existent. The men were so lazy and demoralized that they could not even be bothered to cut fern to lie upon; instead they used the corpses of their comrades for seating and to keep the draughts out from their tents.[11]

Erle's Regiment was little or no better than the rest. An inspection in October produced this report:

> The fact of the Colonel having been ill for some time has not conduced to the benefit of this regiment, because the other officers have not his experience in military matters; so that the regiment is not in a very good state. Nor is it much better than Lord Roscommon's Corps. The clothing is somewhat in disorder, but the Colonel has sent into Scotland for 'surtouts', which are very necessary in this country, and unless the other regiments (Wharton's and Lord Meath's excepted) are supplied with the same kind of clothing, they will not last more than three months.[12]

Already one fifth of the Regiment was sick. By the end of the winter disease and starvation had killed half Schomberg's force. Roscommon's was one of the regiments broken, the survivors posted to make good the gaps in Erle's. From England a draft of 520 men from Luttrell's Regiment was shipped to Ireland to reinforce the sadly depleted units. It was fortunate that the condition of King James's army was on a par with Schomberg's and, although it sometimes harassed its opponents, it was no serious threat during that terrible winter.

Profiting from the lessons learned the previous year, King William proceeded to collect a further force of 27,000 men from the Low Countries and England for the 1690 campaign. Dutch, Danish, German and English, most of the units were well seasoned. By May ships were arriving daily at Dunkald with supplies and reinforcements, and by the end of June King William himself took the field with some 36,000 men, the number including those who had been rotting at Dunkald and elsewhere the previous winter.

Two weeks after King William landed, his army met that of King James at the River Boyne. The next morning the King with his cavalry crossed at Dunmore, while the main body of infantry attacked in the centre. Meanwhile a force of some 10,000 infantry, including Erle's Regiment, moved upstream to ford the river at Rosnaree. Here the crossing was disputed by two regiments of King James's dragoons and when the infantry succeeded in reaching the enemy bank the attacking regiments found themselves in a bog such as few of the men had previously experienced; to further impede them the countryside was intersected with drainage ditches and high hedges. Nevertheless, the attack succeeded all along the line, despite stubborn fighting by some of the Catholic troops. In the end King James's forces either broke or retreated, losing between 1,000 and 1,500 men killed, including stragglers who, when caught, were slaughtered without mercy. Fighting in Ireland has always been savage. King William lost only 400 men, but among them was the gallant 82-year-old Marshal Schomberg, a sad loss to the Protestant cause.

King James himself fought bravely and was wounded, but then fled the battlefield, deserting his followers to take ship for France. Small in scale though the fighting had been, the fate of both Ireland and Europe had, for the time being, hung in the balance. It had been a continental war in miniature, with French troops and generals on the Catholic side fighting a Protestant confederation of European powers. In the aftermath of the

'Glorious Revolution', William's new-found throne had tottered; both the Church and Army had been disaffected, and half the men in public life, fearing a restoration, had been in secret communication with the Jacobites. The victory on the Boyne removed the threat of counter-revolution and restored confidence in William.

The new regiments, including Erle's, had seen proper action for the first time and had learned something of their trade. As all Green Howards well know, some from bitter experience, the Orange Lodges of Northern Ireland (their name taken from the Prince of Orange) after three centuries still celebrate this victory each year on 12 July, bearing at the front of their parades their colourful banners depicting 'King Billy' crossing the 'Boyne Water'.

When King William reviewed his victorious army outside Dublin on 7 and 8 July, the strongest English regiment was Erle's with 693 'private men' paraded. But although the decisive battle of the war had been fought, eighteen months of hard campaigning still lay ahead. The French regiments remained to fight on; although Dublin had fallen, the Catholics were to conduct an effective retreat towards the south-west of the island.

As might have been expected with troops not yet completely disciplined, plundering was a serious problem after the Boyne victory. Orders against it were utterly ignored, so moving the Reverend George Story to utter the perceptive comment that 'it were better that good Rules were not made, than when they are so, they should not be observed and the breakers escape punishment'.[13]

After detaching a strong body of troops to besiege Athlone, an operation which proved unsuccessful, King William concentrated the whole of his army outside Limerick, then Ireland's second city, in which the larger part of the Irish Catholic forces had taken refuge. The siege began badly enough, a body of Irish cavalry surprising the encamped English artillery by night as it moved towards the city, and destroying the larger part of it. Erle's Regiment formed part of the army's advanced guard as it approached Limerick on 9 August. During a pause in this advance, some of these English soldiers were seen to sit down and discuss the likelihood of their bread arriving. This produced criticism from some serious-minded Danish troops, drawn up, as Strong describes, on the English left with 'all the care and circumspection in the world' and who concluded that their allies had no stomach for the coming fight. Whether these English were members of Erle's we do not know, but the Danes had misjudged them. When the advance began again, the Irish loosed a volley from behind a hedge, which awakened a cry from Erle's of 'Ah ye Toads, are ye there, we'll be with you presently'. Then, led by their Colonel, the Regiment charged across the field, directly at the hedge. And so, fighting in this way from hedge to hedge, they and the Danes drove the Irish back to the walls of Limerick for the loss of less than a dozen men.[14] The previous winter, outside Dunkald, it had been remarked that 'We Englishmen will fight, but we do not love to work'.[15]

King William was not to take Limerick that summer. His siege artillery had been largely destroyed in that enemy sortie and the Catholic Irish were now fighting with the courage of desperation. When they made their main assault on 27 August the Protestants penetrated the city walls, but were driven out with the loss of 500 killed and 1,100 wounded. Whether Erle's took part in this attack is not known. But with the autumnal rains now falling in torrents, the siege trenches waterlogged and powder running short, the King decided to withdraw into winter quarters so as to avoid a repetition of that disease-ridden winter at Dunkald.

Serious problems awaited King William in England, and there he took himself, leaving the Dutchman, Lieutenant- General Douglas Ginckel, to command in his place. At the same time a subsidiary expedition under John Churchill, newly created Earl of Marlborough, sailed from England for the south-west of Ireland, where it captured in rapid succession during September both Cork and Kinsale. Within five weeks he was back in England. 'No officer living,' declared King William, 'who has seen so little service as my Lord Marlborough, is so fit for great commands.'[16]

During the winter and spring of 1690-1 Erle's Regiment took part in a number of small expeditions, preludes to the main summer campaign which opened in June, after new clothing and equipment had reached the army from England. Ginckel took Ballymore in West Meath, garrisoned by 1,000 men, within twenty-four hours, after which he moved on Athlone. Although the town walls were quickly breached, the Catholics resisted bitterly, flinging back successive assaults into the River Shannon which ran beneath the ramparts. Eventually on 30 June 2,000 men crossed the river by a narrow ford, forced their way into the defensive works and captured the town. Erle's certainly took part in this fighting, few details of which are recorded.

Ten days later Ginckel marched out of Athlone to encounter the Catholic army on 12 July at Aughrim in a strong position fronted by a bog backed by hedges and ditches. In a subsidiary attack four regiments of foot, one of which was Erle's, were ordered to cross the bog, through which ran a deep rivulet, and drive the Irish from the hedges on the far side.

Erle's advanced first, with the other regiments following, the men wading through the stream up to their middles in mud and water. Story described what then happened:

> The Irish at their near approach to the Ditches, fired upon them, but our men contemning all Disadvantages, advanced immediately to the lowest Hedges, and beat the Irish from thence. The enemy however did not retreat far, but posted themselves in the next Ditches before us: which our Men seeing, and disdaining to suffer their Lodging so near us, they would needs beat them from thence also, and so from one Hedge to another, till they got very nigh the Enemies main Battel. But the Irish had so ordered the matter so as to make an easie Passage for their Horse, amongst all those Hedges and Ditches by which means they poured in great numbers both of Horse and Foot upon us.

The English regiments, advancing too impetuously, had fallen into a clever trap laid by their opponents. Shot at from three sides, they fell back in confusion, Erle attempting vainly to rally them with the cry, 'There is no way to come off but to be brave.' It was, declared Story, 'As great an Example of true Courage and Generosity as any Man this day living.' Pursued by the Irish back into the bog, a great many were slain and both Erle and another colonel captured. Somehow Erle escaped but, wounded, was taken once again, to be then rescued by the English cavalry.

Rallied by their general, the broken regiments were halted and reformed to give the advancing Catholics as good as they had received; after three hours' savage fighting, the latter fled, leaving about a third of their strength on the ground, including many wounded who were slaughtered where they had fallen.

The outcome of this Aughrim battle moved Story to observe that Englishmen were: 'commonly fiercer, and bolder, after being repulsed than before; and what blunts the Courage of all other nations, commonly whets theirs. I mean the killing of their Fellow Soldiers before their Faces.'[17]

And many of their fellow soldiers had been killed. Losses in Erle's that day were to be greater than in any other regiment, four officers and eighty-seven killed; in addition nine officers, including their Colonel, and seventy soldiers were wounded.[18] The Regiment had, for the first time, known severe fighting.

On 25 August Limerick was besieged once again, but operations went slowly, primarily because of the wet Irish weather. In the end, and after an assault in which the grenadiers of all the regiments engaged took part, the garrison blew a parley and surrender terms were arranged. The French troops were permitted to return to their own country, and with them went most of the Irish Catholics, to form the famous Irish Brigade or 'Wild Geese', allies of France and enemies of Britain on many future battlefields.

Ireland was firmly in William's hands and the war was over. So, early in 1692 Erle's Regiment, by then brought up to strength again, was able to return to England. Probably because of his wounds, its Colonel had preceded it the previous summer, carrying despatches addressed to the Queen.

★★★

Eighteen months earlier, in July 1690, Colonel Francis Luttrell had died at Plymouth, where his Regiment was then stationed, at the early age of thirty-one. His brother, Captain Alexander Luttrell, serving in the Regiment, had every reason to expect that he would succeed to the command, much family money having been spent on it, but instead it went to Colonel Thomas Erle, who now commanded the two Regiments, soon to be amalgamated into one of two battalions, both for the time being stationed at Plymouth. Alexander Luttrell and several other officers resigned their commissions in disgust, but the new Colonel, as we have already seen, had been well chosen, and already launched upon what was to be a distinguished career.

Something should be said about the organization of a battalion of foot at this time (regiment and battalion were often synonymous terms). This is clearly set out in the orders for the raising of Erle's own unit, issued from the Court at Whitehall on 16 March 1689. They stated that the Regiment was to consist of thirteen companies, each of sixty private soldiers, three sergeants, three corporals and two drummers (the thirteenth company would have been grenadiers, armed with the new hand-grenade, a kind of small shell, lit by a fuze and thrown in the assault). As well as the colonel and a lieutenant-colonel, who commanded in the absence of the colonel, the officers included a major, an adjutant, a chaplain, a surgeon, an assistant surgeon and a quartermaster, together with a captain, a lieutenant and an ensign for each company.[19] Establishments of this type did often change, ten or twelve company battalions being more usual, and units were, of course, rarely up to strength.

The regiment was, in effect, the property of its colonel, a source both of prestige and income. As with Luttrell and Erle, the monarch might request a loyal and distinguished subject to raise a regiment when he needed extra troops. On the other hand a colonel could sell his regiment, as an officer could sell his commission. This odd system, customary in nearly all armies, required a young man to buy his first cornet's or ensign's commission; thereafter, if he had the money, steps in rank would be purchased from those senior to him, free promotion being granted only to fill vacancies on active service. Military ability or potential were not often factors to be considered. Promotion from the ranks was rare, but not quite so rare as has often been thought. In April 1691 the adjutant of Luttrell's was a promoted sergeant, and early in the 18th century as many as thirty sergeants were granted commissions as ensigns for service in North America, an unpopular station.[20]

But to revert to the regiment. A colonel, as we have seen, needed to be monied in order to be able to clothe, equip and pay his troops until such time as he was repaid by the state – a repayment more often than not delayed. Because of this both officers and other ranks would have to wait for their pay, a serious matter, especially for the private soldier at a time when rations and forage were not even provided on active service but bought out of his meagre pay of eight pence daily from the sutlers who followed the troops. Much of Colonel Erle's surviving correspondence held in the archives of Churchill College, Oxford, concerns the problems of obtaining the money due to him through the chaotic financial channels of the day. One of the consequences of such a system was widespread corruption at every level, a vice in any case common at this time when public office of any sort was seen as an opportunity for profit. Every sort of racket was rife from the holding of imaginary men on roster rolls to outright defalcation: many officers at every level tried to defraud the government and often their soldiers as well, with the government in its turn robbing both the officers and the soldiers.

It is extraordinary, therefore, that despite such problems, these British regiments were starting to demonstrate a gallantry and ability in battle that would arouse the respect of foreigners, both allies and enemies. Better discipline and morale had much to do with this, but their equipment, tactics and field administration were also being steadily improved.

The raising of Luttrell's and Erle's Regiments had coincided with the virtual disappearance of the pike, the invention of the bayonet having made it redundant. When exactly the pike was discarded we do not know, but a statement of strengths and weapons prepared by the lieutenant-colonel of

Erle's in 1691 shows that each company was equipped with eleven firelocks (capable of firing two shots each minute when well handled) and forty-four of the old matchlocks, and that there were no pikes. The battalion suffered from a shortage of 'Bagonetts' and none were very good;[21] these bayonets were most probably the early plug type, rammed into the muzzle, so preventing the weapons from being fired so long as they were in place. The socket bayonets, screwed on to the barrel, were brought into general use two years or so later. Nevertheless, although the French had invented this new method of attaching bayonet to musket, both the plug bayonet and the pike were to remain in use in parts of the French army long after the British had discarded both them and the slow-firing matchlock, the last named also retained by the French for some time to come. The weapons might be sound. Clothing was another matter. In this same report it was described as 'enough for our Complement of men, but they are such as will hardly hang to their backs being soe thyn and shattered'.

★★★

King Louis XIV of France reacted to William's seizure of the English throne, not just by supporting James's invasion of Ireland but also by declaring war against the Dutch. When King William's victory at the Boyne freed him to take personal command of the Allied forces on the continent in 1691, the campaign at first went badly. This impelled Parliament to vote the funds to despatch 23,000 British troops overseas in 1692, among them the battalion raised by Luttrell. (For clarity it will now be referred to as the 1st Battalion of the Regiment, and that raised by Erle which fought in Ireland as the 2nd Battalion.) Originally the 2nd Battalion was to have been sent, but this was later changed, Erle being ordered to embark 'thirteen companies of the Two Batall' under his command. It seems that ten companies came from the 1st Battalion and three were found by the 2nd.[22]

A sound start was made by this inexperienced 1st Battalion, none of whose members had seen action other than the few who had served in Ireland. After landing at Ostend it marched towards Bruges. The French laid an ambush for Erle's men, but the English somehow learned of this and managed to turn the tables. There are two different accounts of the outcome. In the first, forty of the enemy were said to have been killed and the same number wounded. The other states that of the French force of 4,000 horse, with foot as well, 150 were killed, the French commander being numbered among the prisoners; many horses were taken and sold, the money being distributed among the soldiers.[23] Whichever account is correct, the raw unit had acquitted itself ably.

Warfare at that time was an involved matter, its various manoeuvres often hard to unravel. As is usually the case, strategy was dictated by logistics, to such an extent that there were severe limitations on the type of terrain over which armies could be deployed. With rudimentary or non-existent roads, allied to supply systems often ill organized and ineffective, armies largely depended on what they could buy or loot, and on magazines being prepared in advance of the campaign. For this reason the Spanish

Netherlands (what is now Belgium) over the years suffered far more than their share of the horrors of war. Its strategic position was unfortunate, lying as it did between France and the Protestant Dutch Netherlands; to the east was the vast and complicated agglomeration of states owing allegiance to the Emperor in Vienna – what is now Germany, Austria and part of Italy. And, across the Channel, lay England, a country reluctant to allow an unfriendly power to control the ports facing London. It was, however, the rich agricultural produce and highly developed system of rivers and connecting canals, over which both troops and stores could be transported, that made the Netherlands the cockpit of Europe, 'a country where men could kill each other without being starved', as that fine military historian, Sir John Fortescue, so eloquently put it.[24]

The major rivers, among them the Lys, the Scheldt, the Sambre and the Meuse, all ran in a north-easterly direction, natural routes for French penetration towards the Dutch Netherlands and the Empire. To protect these waterways, fortresses had been built by Vauban and his brilliant school of military engineers: rings of massive earthworks, constructed in depth and with sally ports for sudden forays against attackers, were protected by cleverly sited enfilade fire. Because the cumbrous artillery of the day was difficult to move and short in range, its effect against such defences was limited. Sieges, therefore, were often protracted and laborious; the attacker having to dig approach and parallel trenches and battery positions. Perhaps the onset of winter might force the attacker into winter quarters. Perhaps the defenders' field army might oblige the attacker to raise the siege. Failing this, one of two things might happen. The fortress commander could surrender in good time, possibly marching out with the honours of war; the alternative was for the garrison to risk massacre and the city the sack if the attacker were forced to storm the fortress with all the consequent loss of life.

Warfare tended then to be slow moving and formal, with pitched battles avoided unless conditions were unusually favourable. Armies and their equipment were expensive and hard to replace. Elaborate manoeuvres were often attempted to threaten an enemy's communications and compel him to retire from important areas, but it took an outstanding commander to fight such a war of movement and win battles cheaply.

King William, although competent in many ways and conspicuously brave, was not such a commander. Moreover he was the commander-in-chief in a discordant coalition, his regiments provided by a variety of states, great and small, some unreliable and all self-seeking. And he was usually outnumbered by the best organized and equipped army in Europe, one controlled by a single individual, King Louis XIV. The consequence for the Allies was a series of major setbacks and outright defeats in several of which the 1st Battalion of the Regiment was to be involved.

On 5 June 1692 Namur, one of the most important fortresses on the River Meuse, had fallen to the French. So as to draw William's army away from the city and prevent its recapture, a French army under Marshal François de Luxembourg moved as if to threaten Brussels.

Twenty miles to the south-west of that city, the two rivals met at the village of Steenkerke on 3 August 1692 in what contemporaries described as the bloodiest battle ever fought by infantry alone. Unfortunately regiments were rarely mentioned by name in despatches, but it is almost certain that the 1st Battalion, fresh from its success outside Bruges, was a part of the Allied army.

King William almost succeeded in surprising the French, but his tactical handling of his troops was ill managed. With twelve battalions, British and Danish, in the lead, the Allies attacked and broke through the first French line, but this advanced guard took heavy losses from the withering fire of the French infantry, posted behind hedgerows on higher ground. But when his second and third lines began to falter, Luxembourg threw in his reserves, all picked regiments. This was too much for the already heavily hit and outnumbered British and Danes, who were forced back. The ground was too broken for the Allied cavalry and the supporting infantry failed to arrive, the cavalry impeding its movement. In the rear all was soon confusion. With the battle clearly lost, King William withdrew, individual British and Dutch units covering the retreat with great gallantry; among them were the grenadier companies of the British regiments, one of which would have been Erle's.

Steenkerke marked the end of the 1692 campaign. It also showed that the British Army was now a force to be respected, its regiments both aggressive and stubborn. The dead on each side numbered some 3,000 with the same number wounded, the losses among officers being especially heavy.

Some indication of the part played by Erle's during the summer's fighting is revealed from a casualty return that covers the period from June to September. Present at the end of the summer were a mere 169 men, 116 being absent elsewhere. During the campaign, twenty-one men had deserted, five had died, one was a prisoner, twenty had been killed and seven wounded.[25]

★★★

That November, Colonel Erle, still a Member of Parliament, made his only recorded speech in the House during the debate on the employment of foreign generals in the British service. His moral courage matching his physical bravery, he criticized his king's handling of the battle, especially blaming him for not obtaining more accurate information about the ground over which it was fought. Erle was in every way an officer of character. In Ireland his general had described him as 'a man of very good sense, a hearty lover of his country and likewise of his bottle'.[26] He appears to have been a small man, for his boots (which still exist) measure only 9 inches externally. Irritation often flashes through his letters and he seems not to have suffered fools easily.[27] His ability as an organizer and commander matched his bravery in action. He was to have a long and active career as one of Marlborough's most trusted generals.

For the 1693 campaign Erle received his first step in promotion as commander of the brigade in which his 1st Battalion was to serve. Bad weather delayed the opening manoeuvres until the beginning of May, but the two armies then found themselves facing each other in strong positions, neither daring to attack, both ill supplied and discontented, and both losing large numbers of men through desertion.

Luxembourg was the first to move, and in an easterly direction, to lay siege to the fortresses along the Meuse. As he had hoped, he drew William away from his impregnable camp in an attempt to relieve them; at the same time the Allied commander detached forces to reinforce these Meuse garrisons and to make a diversion along the French lines of the Scheldt and Lys. Luxembourg had created the opportunity to attack a much weakened enemy.

William had entrenched himself strongly in a position partly protected by the marshy Landen Beck, its right resting on Neerwinden village and its left on Neerlanden, four miles away. Just to the south of the latter lay Rumsdorp village, which William had converted to a strongpoint, a little in advance of his main defence line. To defend this position William could count 50,000 troops. Luxembourg mustered 80,000. Erle's brigade, which contained his 1st Battalion, garrisoned Rumsdorp. The brigadier himself had been sick with a fever, but hearing that action was imminent, he mounted his horse and managed to reach his brigade in time for the battle.

Early on the morning of 19 July the French attacked, their main thrust being against Neerwinden, held by battalions of British and Dutch Guards and some Hanoverian units. After very savage fighting, this assault failed. Rumsdorp, garrisoned by the 3,000 men of Erle's brigade, was then assaulted by a French force more than four times its size. The fighting among the village houses and hedgerows matched in intensity that at Neerwinden, but the outnumbered defenders were in the end thrust out; Erle is recorded as having behaved himself very gallantly before being seriously wounded.[28] When reinforced, however, his units rallied and recaptured part of the village.

Meanwhile Luxembourg was making a second attempt against the Allied right. This also failed, but French reserves were plentiful and the battered Allied survivors were nearly out of ammunition. Neerwinden was lost and the line began to crumble. Despite repeated charges by the English cavalry, on this right flank retreat turned into rout. Bridges across a stream lying behind the British front were soon blocked by a mass of guns, wagons and men. Thousands of fugitives were cut down or trampled to death. On the Allied left, however, the withdrawal went steadily and was marked by a desperate attempt by Erle's Regiment to save the guns.

Casualties at Landen were heavy indeed: as ever, figures differ, but one-fifth of the Allied and one-tenth of the French army seem to have been killed or wounded; the nineteen British regiments lost 135 officers. In Neerwinden alone, 5,000 bodies were said to have been counted the following day. The next summer a thick carpet of crimson poppies spread across the battlefield, flowers, the peasants said, that had been watered by the blood of the slain.

In a letter written a short time after the battle, an officer suggested that although all the heavy cannon had been lost, and the French had been left with the honours, 'we don't find ourselves half as much beaten as at first we

thought we were' but that 'we have all now had our bellies full of fighting'.
The writer's complaint that 'some of the English horse made as much haste
to preserve their dear persons as anybody there' suggests that he was most
probably an infantryman.[29]

The campaign of 1694 was uneventful. The 1st Battalion was still a part
of their Colonel's brigade, and the two armies marched and counter-marched
without joining battle. That winter the army built for itself 'barraques' or
huts of straw, a word that was just coming into use to describe soldiers'
accommodation. In such conditions fire was the great dread, and when in
September a Frenchman was taken with a lighted match near the
ammunition wagons, his punishment was to be burnt alive. Sabotage was
not yet treated as a legitimate act of war.[30]

By 1695 the expenses and losses of the continual fighting were starting
to tell upon the French; also far too high a proportion of their forces had
become immured in their fortified lines which, because of their previous
successes, now extended in a line from Namur to the sea. This was a strategic
error which the French were to repeat during the subsequent centuries, a
habit noticed by Fortescue at the end of the 19th century.[31]

King William seized his chance. Masking his designs by a series of
feints, he marched to Namur, Vauban's masterpiece, standing at the junction
of the Meuse and the Sambre. His first, and successful, assault against the
outworks took place on 6 June. Casualties were heavy, as they were during
the two subsequent attacks on the town and the final assault on the citadel,
in all of which British regiments saw hard fighting. Erle and his brigade of
six battalions formed part of the covering force, and as such saw little action
but plenty of leg work; regiments from this covering force were used to
relieve others taking part in the siege, but details do not exist. Erle's own
grenadier company, however, was rather more closely involved, eighteen
members taking part in the initial assault against the outworks, and the whole
company, together with the grenadiers of all the regiments of the besieging
force, in the final victory on 20 August. Of all the regiments, losses among
Erle's grenadiers were the second highest.

★★★

The following year, twenty battalions, among them Erle's, were rushed
back to England to cope with a threatened French invasion that was linked
to a conspiracy to assassinate King William. For five of its companies the
voyage was eventful. Together with a battalion of the Third Regiment of
Foot Guards, they were captured by French privateers; the following year,
however, they were to be released when the Treaty of Ryswick was
concluded,[32] victories such as Landen having almost ruined the French. This
temporary peace marked the end of the first stage of that long-drawn-out
conflict between Britain and France. For the first time since the Hundred
Years War, a large British army had fought in a series of protracted
campaigns on the continent of Europe. It had proved itself the equal or better
of any of its allies, a force far removed from the rabble that had starved
during that winter at Dunkald.

But, as always, Parliament was now reluctant to vote the money needed to keep in being large forces in peacetime. The old cry of 'No Standing Army!' was quickly raised; once again voices declared that the Militia was all the country needed for its defence. Not for the last time, the Regiment's 2nd Battalion disappeared – disbanded in September 1697.

Since the Irish War the 2nd Battalion had only once seen action, in an incompetently organized combined operation against Brest, the purpose of which was to attack the French fleet in harbour, it having proved impossible either to bring it to action or adequately to blockade it. But when intelligence reached the French that such an operation was in hand, they reinforced the base. Again the exact part played by the 2nd Battalion in what was to prove a disastrous operation is not known. The landing from the ships' boats, led by the brigaded grenadier companies of the 7,000-strong force, commanded by Brigadier-General Lord Cutts, was ill organized, covering fire was inadequate and the defences proved to be strong and well manned. The attack failed and British casualties were understandably heavy.[33] A postscript to the affair is provided by the report on an inspection of the Battalion at Salisbury carried out by Cutts in the following January. At this time morale among the home-based units was low, especially among those anticipating overseas service. Cutts wrote:

> His Majesty will see the number [now lost] and be pleased to consider that this was one of those regiments ruined at sea the last summer, particularly the grendiers [sic] were almost all killed and made prisoner. The men I saw were . . . good, seasoned men, and though they were ill in clothes . . . yet they appeared neat and clean. In the whole I really think it is a good battalion. I enquired particularly about their Irish papists but they tell me they are all discharged.[33]

Only thirty line battalions survived the war's end, three on the English establishment, six on the Scottish and twenty-one on the Irish. Among the latter was Erles's remaining battalion, and to Ireland it moved in 1699 at the much reduced establishment of eleven companies, each of thirty-six private men, two sergeants, two corporals and a drummer. There were eleven captains, the Royal Proclamation specifying 'whereof the said Colonel, Lieutenant Colonel and Major were three'. As was the custom, these senior officers each commanded one of the line companies and drew the pay for that company, the actual duties of command being delegated to a lieutenant, the one in charge of the colonel's company being designated 'captain-lieutenant'. Otherwise the officer establishment was unchanged.[34]

As for Erle, his king was recognizing his worth. In 1696 this stalwart soldier was promoted to the rank of major-general and appointed Governor of Plymouth. Although he had served in action with his Regiment for the last time, his interest in them, financial and otherwise, was undiminished, as his papers reveal.

Above: 'The Raising of the Regiment' at Dunster Castle. Oil painting by Terence Cuneo. It represents Colonel Francis Luttrell supervising the attestation of his men outside the main gatehouse of Dunster Castle, the only external part of the building still surviving. The men are dressed in the Luttrell livery of blue lined with yellow, and are shown being issued with their muskets, bayonets and pikes. (Reproduced by courtesy of the artist)

Right: A showcase at the Museum displaying one of the muskets with which the Regiment was equipped in 1688. To the left is a very rare set of 'Twelve Apostles', separate wooden containers holding twelve powder charges.

The Raising of the Regiment 1688

1688 Why Luttrell's Regiment of Foot Soldiers was formed

Left: General the Right Honourable Thomas Erle. Copy of a painting by Sir Godfrey Kneller.

Right: Colonel Francis Luttrell. Copy of a painting in Dunster Castle.

2. THE FRENCH
WARS: 1697-1793

The peace that followed the Treaty of Ryswick was short lived, no more than an uneasy four-year interlude. In 1701 there began the War of the Spanish Succession, the result of the Spanish throne passing to a grandson of Louis XIV's and the consequent threat to the Spanish Netherlands. In Britain enthusiasm for the coming war, both in Parliament and the country, was inflated by King Louis's pointless arrogance in proclaiming King James II's son as King of England, Ireland and Scotland when the old King died in Paris in the September of that year.

With the British and Dutch navies holding command of the sea, men and supplies could be moved freely across the Channel and the North Sea. Elsewhere the balance of power had, at the outbreak of war, shifted smartly towards the French, their armies now in control of the whole of their new Spanish inheritance. This included not just Spain itself, but much of Italy and the contested fortresses of the Spanish Netherlands. Moreover the Roman Catholic state of Bavaria was also allied with England's enemy.

When King William died in a riding accident in 1702 he was succeeded by the Princess Anne, the Protestant younger daughter of James II, but British policy towards Europe in no way changed. Already, in 1701, Marlborough had taken a dozen battalions to Holland. In the following year there ensued the customary scramble to make serviceable what little remained of the country's neglected army and raise new units to replace those so wantonly disbanded.

At the start of war, Erle's Regiment was still in Ireland, as was its Colonel, now the Commander-in-Chief there. His letters reveal his many problems in finding the officers, recruits, serviceable weapons and equipment needed to bring his Regiment up to war strength. In this voluminous correspondence, all conducted in his own hand, seemingly without the benefit of clerical help, his frustration and exasperation become clear from this single extract:

> Confound ye Capt. Mead, why does he not come to his post, if I could catch him after the Regiment is gone, I'll certainly shoot him for a Deserter. We shott a poor devil of a private soldier of Charlemont's Regiment last week. I am sure an officer that is wanting on this occasion deserves it, much better.[1]

The 'occasion' was a seaborne attack against the Spanish port of Cadiz, which sailed towards the end of July 1702. Erle had hoped to be put in charge of this expedition (at the same time he could have kept an eye on his Regiment),

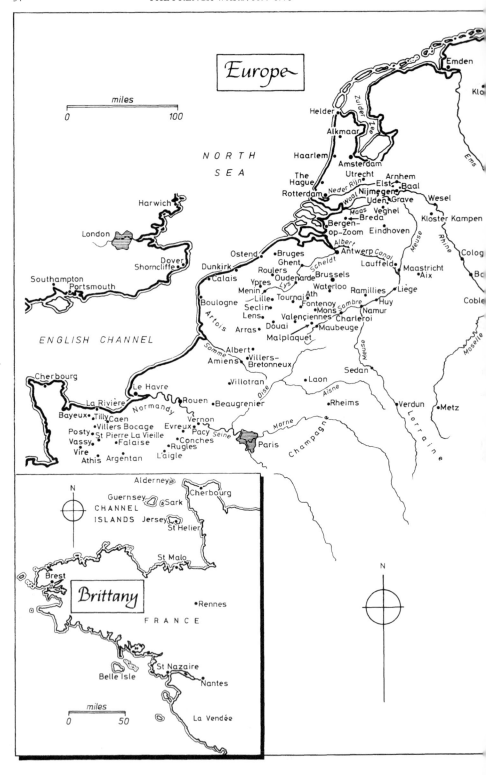

but the command fell to the incompetent Duke of Ormonde, which was a pity because Erle would have been a far better choice. Half of the dozen British regiments in the force had been raised only that year, so Ormonde was probably being honest when he told Erle that he 'was glad that your regiment goes, believing it to be the best'. After the expedition had sailed Erle also learned that 'My Ld. of Ormonde sends me word he has liked ye Army very well: which I think you would be glad to know'.[2]

The expedition was an utter failure. Ormonde and the admiral could not work together, English quarrelled with Dutch and sailor with soldier. When the fleet arrived off Cadiz in the middle of August the Spanish batteries repulsed with heavy loss an attack against the fort protecting the harbour. Another body of Allied troops then occupied the adjacent towns of Rota and Port St. Mary, which they looted with enthusiasm, despite strict orders to the contrary. Even churches were plundered and despoiled. Because Erle's Regiment was so highly thought of by Ormonde, it is likely that it took part in the main assault, but no details are known. In the end the expedition did no more than further alienate Spanish sympathizers.

Erle himself did not favour such small seaborne forays.[3] He was not alone in this. From the Tudors right through to the present century it has often been argued that small bodies of troops, helped by the Royal Navy, can bring a decisive influence to bear upon an enemy, forcing him to disperse his forces to guard against such threats; it was a temptingly cheap way to achieve victory, an additional bonus being the capture of commercially valuable colonies. On the other hand, to place a large land army alongside those of allies fighting a continental campaign was expensive in lives and could become an open-ended liability, its price in men and treasure ever escalating. But, as the successors to Erle's men at Cadiz would learn to their cost on many bloody shores, this 'blue water' or maritime strategy would produce many more failures than successes.

***

During the course of its ignominious return from Cadiz, Ormonde's expedition was consoled by the capture of the Spanish treasure fleet returning from the New World. The booty amounted to a million pounds, but Erle's Regiment gained no share in the division of the spoils. With four other units it had been shipped direct to Jamaica, the starting-point for an expedition against Guadaloupe in the Leeward Islands, the richest of the French sugar islands.

The composite assault party which landed there by boat on 12 March 1703 consisted of a major, two subalterns and 220 men from Erle's, together with a similar number from each of the other four regiments – for tough assignments units were very frequently intermixed in this way, regiments usually but not always having to surrender their grenadier companies. The fight to get ashore was sharp, but the landing was quickly made good, so forcing the French to retire into their citadel of Basse Terre. The British commander, lacking artillery, then asked the admiral for ships' guns to bombard the fort, but met with a refusal because of the danger that the

French fleet might intervene. In the end, however, the French blew up the castle and retired inland.

For the besiegers it had been an unpleasant and costly business. There were no tents, and the meals of porridge and peas, sent ashore already cooked, were always eaten cold. Bread ran short in April and half-rations resulted. In such conditions the death toll mounted.

After the French withdrawal the British systematically destroyed their coastal settlements and fortifications, driving their troops into the forests, where they must have suffered in much the same way as their attackers. In May, however, the British relinquished their gains and left, taking with them the captured French guns. It was done in the nick of time as a French relief expedition from Martinique was already at sea.

The strength of Erle's Regiment when it embarked from Guadaloupe was only eleven officers and 309 other ranks. Casualties during the fighting had been high; three officers and sixty-eight men killed and wounded. Another officer and twelve men were reported to have died of disease. These figures indicate that the total mortality since the Regiment left England had been about eighteen officers and 300 men; if this is so either losses at Cadiz had been high or a large number of men had died at sea,[4] a not unusual occurrence. Even in the present century conditions in troopships have often been squalid and unhealthy: in 1703 transports were floating pesthouses, a private soldier writing of 'continual destruction in the foretop, the pox above board, the plague between decks, and the devil at the helm' with men 'labouring under many inconveniences, having only the bare Deck to lie upon, which hardship caused abundance of our men to bid adieu to the world'.[5]

So what was the object of this tip and run raid, except to inflict temporary hurt upon the French? No sound answer has come to light. It all appears to have been a waste of men and treasure.

The Regiment is rightly proud of its Battle Honours, but their award has often seemed capricious. Since the First World War they have been granted without distinction between victories and defeats: Gallipoli, Norway and Dunkirk are inscribed on the Colours, but not Landen. Five times during the French wars the British conquered Guadaloupe, only to quit the island afterwards, as in 1703, or hand it back to the French at the subsequent peace treaty. Twice, for the 1759 and 1803 expeditions, Guadaloupe was awarded as a Battle Honour even though the fighting on both occasions was far less intense and casualties far fewer than in 1703.

★★★

Details of the Regiment's next expedition are better recorded and make even unhappier reading. Together with other shattered regiments from various West Indian garrisons, it was shipped to Placentia in Newfoundland, then a French colony, with the aim of disrupting the valuable trade in fish and timber. When the fleet arrived off the island the ships were enshrouded in a fog that made navigation impossible for thirty days, so that the soldiers rotted aboard, unable to withstand the damp and bitter cold of the North

Atlantic after the sweltering heat of the West Indies. They were, a chronicler relates, 'benumbed in their limbs and subject to fluxes and scurvies'.[6] Five thousand soldiers had embarked; soon only 1,038 remained alive. There was nothing to be done but bring the survivors home.

Stationed first in Ireland, the Regiment's problem was to find the recruits needed once again to bring it up to strength. To make good casualties on such a scale and to man the many new regiments then being raised was near impossible. Only the unemployed could be conscripted; there was a limit to the number of gullible or drunk labourers into whose pocket the Queen's shilling might be slipped; condemned men could escape the gallows by enlisting and the bankrupt the debtor's prison. How sufficient recruits were found is hard to imagine, but they were.

Since 1690 regiments had been recruiting their own men. So one can detect the satisfaction in the letter Captain George Freke wrote in May 1704 from Waterford to his Colonel, now promoted to lieutenant-general and about to become Master General of the Ordnance. His news was that, 'Capt. Cradock landed here yesterday. He brought over with him 50 recruits, eighteen of which are for his own company. Extraordinary good men . . . they are all good men – such as I am very sure your excellency will approve of'.[7]

The following year Erle used his influence to move his Regiment to join him in England. Stationed at Chester on garrison duties, its companies were scattered all over the north of England, with Erle obliged to write most apologetically to a Mr Robinson, apparently the Lord Mayor of York, that he was 'sorry to understand that there should be the least occasion of complaint of any disorders and must impute it to the negligence of the officers now at the head of them'. Cradock, now a major, was to be posted immediately to York in place of the previous detachment commander 'whose indisposition obliged [him] to quit the Service'. Erle mentioned also that he would have preferred another company to have been quartered in the City because of the difficulty of disciplining these outlying detachments.[8] How Erle dealt with the offenders is not known, but the year before he had issued this sarcastic rebuke to an erring captain who had been enjoying himself in London and Dublin.

> I find that you have gone through great hazards, and hardships thys winter, so that truly I believe you could hardly have endured if you had been doing your duty with yr company, for want of which I cant expect an account of their condition from you, but from other hands I understand that it is in a very ordinary way. By the industry of the major and the diligence of the few officers I have had with me (we having never been off the Horseback in order to do it) I hope men will not be wanting, though they will prove very chargeable, but the Arms and Accoutrements I expect you will take care to see complete by that time'.[9]

There followed a threat of a personal interview and the sack. Again we see Erle's close interest in the small details of his Regiment: not much seems to have been left to the lieutenant-colonel.

This tour appears to have been the start of the Regiment's long association with Yorkshire. Was there some connection between this visit and the despatch of recruiting parties from Flanders in 1711? Directed to find seventy recruits from London, fifty from Carlisle, eighty from Yorkshire and thirty from York itself, they seem to have done their job well, for the men they brought back are recorded as having been 'a fine body of recruits'. The recruiters consisted of twelve officers and twelve sergeants, all of whose names are known from Erle's correspondence, a rare if not unique record.[10]

During the Regiment's stay in England there was much talk about the chances of joining Marlborough in Flanders, as there was of Erle himself being posted there.[11] The frustration must have been intense, but in 1708 at last it happened.[12] For the first time the Regiment was to serve under that great commander.

★★★

It was a decade since the Regiment had soldiered on the continent, a period during which the reputation of British arms had been still further enhanced. The credit lay with Marlborough, whom his new monarch had honoured with a dukedom after his victorious campaign of 1702. As Commander-in-Chief of the Allied forces serving in the Low Countries, he had proved himself to be the foremost Captain of his age.

Marlborough was a master of every aspect of warfare. Stressing the vital need to destroy the enemy in battle, he broke away from the stereotyped habits of the time, outmanoeuvring his opponents by quick and unexpected marches, often by night, so forcing them to fight at a disadvantage. His greatest strategical triumph was his 250-mile excursion to the Danube which enabled him to join his forces with those of the Empire and thereby crush the Bavarians and their French allies at the Battle of Blenheim. The risks had been immense: by removing his ninety squadrons of horse and fifty-one battalions, he had left Flanders for a time defenceless; to ensure speed he had discarded all his heavy artillery.

The stern discipline of Marlborough's army was matched by high morale. The latter he fostered by his care for his troops and the victories he gave them, so inspiring them with confidence in a leader, who, like Wellington and Montgomery afterwards, was a familiar figure both on and off the battlefield. Among his tactical innovations, he taught his infantry to fire by platoons, instead of by line as in the past: faster and better-controlled fire resulted. As with Cromwell, his victories were made complete by the shock action of his cavalry – and this at a time when the French were still using cavalry as no more than manoeuvrable fire power.

Like all great commanders Marlborough understood how successful generalship depended upon good administration. Within the limitations of the rudimentary supply system imposed upon him and the appalling communications, he worked near miracles, giving his personal attention to even the smallest detail: half-way on the road to Blenheim, new shoes were ready and waiting; the small cart he introduced was for centuries known to European peasants as the 'Malbruck'. By such methods his units arrived at

the Danube complete and fit, while the French army, endeavouring to
intercept him, lost a third of its strength by desertion and disease.

Although his British battalions and cavalry regiments rarely formed
more than a quarter of his force, their quality was now such that they tended
to be used where the action was fiercest; highly respected by the French,
they set a standard for their German, Danish and Dutch allies to match. It
was a far cry from 1689.

★★★

At Ramillies in Flanders in 1706, Marlborough had achieved another
brilliant victory, a prelude to the Allies driving the French back to their own
frontiers. Antwerp and Ostend both capitulated, thus easing communications
with England: for the first time King Louis made overtures for peace. But
in 1707 the tide again turned against the Allies. In Spain an Allied army, in
which Erle commanded the centre and was once again wounded, was soundly
defeated at Almanza. (In a letter written earlier to an extremely busy Master-
General of the Ordnance, which was either tactless or humorous, or both,
Marlborough had apologized to Erle for 'calling you away from so agreeable
a retirement').[13] Then the Empire treacherously concluded a secret treaty
with the French, so permitting them to reinforce Flanders and the
Rhineland. Faced again by superior numbers and foiled by atrocious
weather, that summer Marlborough could do nothing.

Again in 1708, the French outnumbered the Allies in the Low
Countries, having concentrated even more of their forces there at the expense
of other theatres. After much marching, counter-marching and wide-scale
skirmishing, the two armies at last met near the fortress of Oudenarde that
guarded the Scheldt, twenty miles west of Brussels. To join battle
Marlborough hurried his heavily laden soldiers for fifty miles over the
atrocious roads, a distance they covered in but sixty hours. An encounter
battle, rather than the usual set piece conflict of the time, Oudenarde was a
confused struggle in which Marlborough's fine eye for ground and tactical
skill won the day. By comparison with the French, whose complete baggage
train they captured, Allied casualties were not high. No complete record of
the units engaged has come to light, but it is more than likely that Erle's
was present at this victory. Marlborough would have made full use of such
a fresh and first-class unit.

In the meantime, Erle, having extracted himself from Spain, an ill-run
theatre, had been sailing up and down the coast of Normandy in charge of
a force of eleven regiments tasked with creating a diversion so as to draw off
French troops from the main front – just the type of work he found
unappealing. When the force was eventually ordered into Ostend, bringing
with it valuable supplies, Marlborough made use of Erle to organize his base
and the forward movement and protection of his supply trains, duties similar
to those of a modern base commander and for which Erle displayed much
talent.[14]

The further misfortunes suffered by the French after Oudenarde
persuaded King Louis once again to seek some common ground to end the

war, but the Allies' demands were still set too high, so arousing a mood of
stubborn defiance among all classes of the French people. The result was
that when the Allies stood ready to invade France in June 1709, the French
were ready to fight harder than ever.

After a bitter 2-month-long siege, the fortress of Tournai fell to
Marlborough. Next he threatened Mons, but the main French army, pushed
forward to save the place, was positioned on a plateau near the village of
Malplaquet. As the early morning mists began to clear on 31 August, the
Allies advanced upon the formidable French entrenchments, many concealed
in thick forest. In many ways Malplaquet was similar to the bloody conflicts
of the American Civil War in the next century. Among felled tree trunks,
tangled foliage, streams and quagmires, the confused battle raged, units often
intermixed. At one stage Irish fought Irish, the future Royal Irish Fusiliers
at the throats of the French Royal Regiment of Ireland. The defenders had
the advantage; that morning some 30,000 men were killed and wounded,
about one in five of those engaged, one-third of them Frenchmen. In the end
the French quitted the battlefield, but in good order, the Allies too exhausted
to pursue them.

The Regiment won the Battle Honour of 'Malplaquet', although it
played only a minor part in the fighting. Its name does not appear in the
official order of battle, and only a single contemporary writer mentions it as
being present. It seems likely that it was part of Marlborough's reserve, as
were the Coldstream Guards, also awarded the Battle Honour.[15] In fact, a
letter to Erle from one of his officers exists in which, after recounting the
preliminary moves before the battle, he states 'Your honours Regiment did
not partake of this Honour for [sic] not in the action'.[16]

As always, the Regiment needed a refit at the summer's end, a fact
brought vividly to light by a surviving bill which details every item of the
clothing supplied, near unique evidence of what Marlborough's infantrymen
wore. From it can be calculated the Regiment's exact other rank strength.
The total cost of it all amounted to £1,582.10s.9d. to which was added
£24.18s.0d. for carriage. The first item of 'welch plain red and yellow'
indicates that the facings to the red coats were still yellow (they were changed
to green some time during the next thirty years, probably by accident), the
drummers being the reverse. Sergeants were issued with better quality hose,
cravats, shirts and possibly breeches than those junior to them, while their
cocked hats and their coats were embroidered with silver lace. It is a
fascinating document.[17]

★★★

When the Allied offensive into France was resumed in 1710, the
Regiment was to take part in the siege of Douai which was a major, successful
and well documented battle. Douai has been described as the most bitter
siege fighting in which British troops have been engaged. So strong were the
defences that the attackers spent fifty days in the open trenches, under fire
nearly all the time. The Regiment's killed and wounded numbered 312, more
than half its strength, and more than the entire British casualties at Ramillies.

Sickness must have taken many more, hence that recruiting campaign in Yorkshire the following winter. As late as 1934 the Regiment's request for the award of the Battle Honour was refused.[18]

The next summer's campaign in Flanders was to be Marlborough's last and his most brilliant. During the previous winter the French had constructed a seemingly impregnable defence line stretching from the coast to the River Sambre, a combination of inundations, fortresses and strong field works, a forerunner of the Maginot Line. In the centre the French placed their main army, protected by the strongest of field defences, christened the *non plus ultra*, and deadly indeed to assault. By deception and speed Marlborough outwitted his opponent, drawing his army away from the fortifications, so allowing the Allies to invest Bouchain, the key to a drive towards the heart of France. It was another grim siege, the troops, Erle's Regiment among them, often up to their waists in water. On 13 September the garrison surrendered under the eyes of the French field army, chary of attacking Marlborough's covering works. No record of British losses has yet come to light, but again it was a victory for which a Battle Honour was refused. Casualties in the Regiment must have been high for in the Admissions Register of the Royal Hospital, Chelsea, are inscribed the names of six of its Pensioners wounded at Bouchain, including Sergeant R. Fielding whose left hip had been 'struck out by the springing of a mine' and Private Robert Plumer who was 'Shott in right arm at Doway, left shoulder blown up at Vigo. Hath an ague now'; the names of a dozen of the survivors of Douai are also recorded. In all, thirty-five Regimental Pensioners are listed, but no more than a single one is shown as having fought at any of the other great battles of the time, including Landen, Limerick and Malplaquet, possibly an indication of the scale of the Regiment's losses at those two sieges.[19]

***

One more summer's fighting and the Allies could well have threatened Paris, but while Marlborough had been winning his country's battles on the continent his political enemies had so undermined his position at home that Queen Anne dismissed him from all his offices. The next summer an exhausted Europe concluded the Peace of Utrecht. With it, King Louis's grandson was accepted as King of Spain, but the Spanish Netherlands and the north Italian lands went to the Emperor. Overtly honours were about even, but crippling damage had been done to the economy and political structure of France, whose ambitions to establish a hegemony over Europe had been thwarted for many years to come.

The Regiment was to remain as part of an 8,000-strong garrison in the Netherlands until after Queen Anne's death in 1714, quartered for much of the time again at Bruges. The customary disbandments had followed the signing of the Peace, units often being chosen regardless of their seniority. The Regiment possibly survived merely because it was still abroad, out of sight and out of mind.

In the winter of 1711/12 its official designation changed when Erle sold the Colonelcy to George Freke, now a brigadier-general, who had commanded it with distinction in action since 1709. So ended Erle's twenty-three year connection with his Regiment, so strong a link that for some time to come it continued to be described as 'Erle's'. After the 1708 campaign the old general had returned home, his health injured by repeated attacks of gout, but in 1712 he was appointed a General of Foot. As one of Marlborough's men, he fell from favour with his patron, losing his appointment of Master-General of the Ordnance and Commander-in-Chief of South Britain, but was reinstated again as Master-General in 1714. In the church at Charborough can be seen this inscription:

> In this Vault are interred the remains of Thomas Erle, General and Commanding of the Foot and one of the Lord Justices in Ireland, Governor of Portsmouth, and Lieut.-General of the Ordnance, and one of the Privy Council in England. He raised a regiment in this country at the Revolution which he carried over to Ireland with him and they signalized themselves both at the battle of the Boyne and Aughrim, and afterwards served with great honour both in Flanders and Spain during the reign of Queen Anne. He died in the year 1720.[20]

A fine soldier he certainly was.

Brigadier-General Freke held the Colonelcy for only a short time, Brigadier-General Sutton buying him out in August 1712. Sutton had fought in a great number of Marlborough's major battles including Blenheim, Ramillies, Oudenarde and the siege of Namur. As with Erle, many of his letters have survived, a number of them dealing with officers' promotion and purchase, as well as his own complaints of shortage of money and being passed over for major-general.[21] Possibly because he was hard up, in 1715 he sold the Colonelcy to George Grove, the Regiment's Lieutenant-Colonel, but he assumed the appointment once again in 1729 when Grove died, having fallen from his horse, and held it until his death eight years later. In the end he did make the rank of lieutenant-general.

★★★

Back in England the Regiment was for a short time scattered around a number of towns and villages in the Tilbury-Sheerness area, the men billeted in the usual way in squalid little alehouses, hardly conducive to good order and military discipline. Few barracks existed, and in any case such housing was unpopular, not just because of its cost. In a brutal age, with no proper police, the Army was the only force available to cope with violent industrial strikes, large-scale smuggling, the arrest of dangerous malefactors and even the escorting of criminals to the gallows. It is hardly surprising that they were often hated and feared by large sections of the population. Barracks could be a reminder of the presence of troops, something which many citizen wished to forget. This dislike of barracks was of lesser importance in Ireland where a number had been built in 1707, nor in Scotland where small forts had been provided against the depredations of the Highland clans.

Later in the year 1714 the Regiment moved to the north-east to be spread between Hull and Tynemouth, a four-day journey on horseback when the commanding officer wanted to visit his troops. Again its stay was short for in March 1715 it moved to Ireland, so missing the Jacobite rising of that autumn prompted by the landing in Scotland of King James II's son, later dubbed 'The Old Pretender', with the intent to supplant Queen Anne's successor, her distant cousin, George I, the German Elector of Hanover. Scottish support was forthcoming, but in England the rising scarcely spread beyond the Roman Catholic Northumbrian squires, despite this new king's inevitable unpopularity with his subjects. The few British regiments, some made up of raw recruits, just managed to check the Jacobites. The rising quelled, a frightened Parliament once again augmented the Army, restoring some of the very regiments that had been disbanded two years or so earlier, only to impose further cuts during the two ensuing years.

Spain rather than France had supported the Old Pretender, and in 1718 she mounted a large-scale invasion of Britain in revenge for Vice-Admiral Byng's near destruction of her fleet before war had officially been declared. As with the Armada, tempests dispersed the Spanish ships and only 300 Spanish soldiers managed to land in Kintail, where they were quickly rounded up.

A reprisal against the Spaniards followed in the shape of a raid in September 1719 against Vigo, on the Spanish Atlantic coast, by a force of ten regiments, among which was Grove's, shipped with three others from Ireland for the expedition. The prospects of overseas service were unwelcome. As the troops were due to embark mutiny threatened, a fact hushed up by their commander, Brigadier-General Crofts, who reported to the Secretary at War that 'the men went with great alacrity'.[22] The truth was different; the Secretary to the Lord Lieutenant writing that 'Coll. Grove's was kept on shoar till the others were all shipt, for fear of mutiny, his being the regimt which Brigadr. Crofts thought he could most depend upon to prevent it.'[23]

The Royal Navy landed the troops without difficulty some distance from the city of Vigo which they then occupied. When the citadel surrendered, parties of British soldiers burnt and plundered around the neighbourhood; resistance was slight, casualties from enemy fire numbering only about 300. Disease and the unlimited wine were another matter, one report reading that 'those that fell by the Vine were reported to be the greatest number'.[24] The booty, in muskets, barrels of powder and guns, stocked for a further invasion of Britain, was shipped back to England, and was vast.

The Regiment's part in this not altogether successful foray is not recorded. We know only that after it returned to Ireland in the December, its Colonel learned seemingly by chance that certain of his men, who had been taken prisoner, had arrived back in Spithead and could be collected by an escort; if not, they would be set at liberty.

This small war had one unusual aspect. The British and the French were uneasy allies against Spain.

***

For the next quarter of a century the Regiment alternated between Ireland and Scotland, spending only a small time in England. Extracts from several inspection reports have survived, this one from County Cork, when its strength numbered thirty-five officers and 373 other ranks, a not untypical example:

> In obedience to your Excellency's Command I have viewed the above Regiment. The bodies of the men are very good. The Regt is very well disciplined, performing their Exercises and firing very well. The Cloathing is good, delivered to the men at different times from ye 18th to ye 28 July. The arms and Accoutrments of this regiment are bad. I found the men were regularily heard and no Complaints. The Regiment have tents.
>
> sd. THOS. PEARCE.[25]

Split as the Regiment was into eight detachments, this does not seem at all a bad report.

During this time another change of Colonels was to be of especial significance. Sutton died in 1738. In his place was appointed Colonel the Honourable Charles Howard, second son of the third Earl of Carlisle and an officer of the Coldstream Guards.

In 1744 Howard's Foot was again on active service in Flanders, as was its Colonel, now a major-general and commanding a brigade. One of its last peace stations had been Edinburgh, and among the files of the *Scots Courant* for the years 1739 and 1740 are some fascinating vignettes about the Regiment. To find sound recruits was, as ever, a problem, but we read that 'several bodies of stout, brawny young men, who had been recruited in the Highlands and other parts of the Kingdom. passed through here on their way to their respective corps. The Hon. Colonel Howard's Regiment is so well recruited since it came to this neighbourhood, that 'tis said the companies are above 50 men strong'. Another extract warns that 'The recruiting officers are ordered to be very cautious against enlisting Scots, Jacobites, or Irish Papists'.[26] The Regiment must have been crammed with Scots, but care was required when picking them, the enlistment of Roman Catholics being strictly forbidden at the time.

As ever, drink was at the root of most disciplinary problems. Two of Colonel Howard's unfortunate soldiers fell off Castle Hill when drunk, one of whom 'stept over the counterscarp into the dry mote and broke all his bones'. Another found himself in the city gaol for beating a police officer. Oddly enough the reports were often marked by both humour and a certain sympathy for the soldier.[27]

The ready availability of recruits in Edinburgh in 1739 was in some way due to the outbreak of a trade war with Spain, popular with a nation that remembered the Armada. At the same time yet another continental war was in progress with France, Bavaria and the newly emergent military state of Prussia trying to carve up the Austrian Empire, the Emperor having died and been succeeded by his young daughter, the future Empress Maria Theresa. Again France seemed likely to seize control of the Low Countries, and again a British Army was involved, one that had been raised almost from scratch two years before. At Dettingen in 1743, the last battle in which a British monarch, King George II, commanded his troops in action, they lived up to the reputation they had acquired under Marlborough thirty years earlier. It was this army that Howard's Foot joined the following year.

That summer there was little fighting but much marching. So far as the Regiment was concerned, it was marked by an event of trenchant importance. Serving in Flanders was another regiment commanded by a Colonel Howard, that of Lieutenant-General Thomas Howard. Confusion was inevitable. A method to distinguish between the two was vital. The answer was simple. The more senior regiment, later the 3rd of Foot, wore facings of buff. Those of Charles Howard's had by then changed from yellow to a yellowish-green. So it was that one of these two regiments became 'The Buff Howards', shortened to 'The Buffs' and one 'The Green Howards'. Tradition is tenacious. For close on two centuries this nickname in some way survived, an indication of its general and lasting popularity, to be accepted in 1920 as the Regiment's official title. The first known written reference dates to 1786, when Francis Grose, the antiquarian, who had held a commission in the Regiment, referred to it as such.[28]

***

Moving out from their winter quarters in Ghent in April 1745, the Green Howards formed part of an Allied army made up of British, Dutch, Hanoverian and Austrian detachments, and commanded by the 25-year-old Duke of Cumberland, the son of King George II. Advancing to relieve the fortress of Tournai at the end of April, Cumberland encountered the French army entrenched astride the village of Fontenoy. After driving in the large numbers of French light troops concealed among a mass of small woods and enclosures, the Allies were faced by a smooth unbroken slope leading up to the enemy's main position. On the right were the British and Hanoverian battalions drawn up in two lines, with the Green Howards on the left of the front line.

Ordered into an unimaginative frontal attack, the red-coated British tramped steadily up the half-mile slope, the converging fire of the French guns carving huge gaps in their ranks. Only when they reached the crest could they see the enemy, the pick of the French army, drawn up behind their breastworks. The first enemy volley smashed into them from a distance of no more than thirty yards, but it failed to check that inexorable British advance. Fronts grew narrower as men closed up in the ranks, but still they pressed on, breaking through the first French line and advancing for a full 300 yards into the heart of their camp. Halting then to deal with the French reserves, the British with superb fire discipline brought their muskets to their shoulders to fire successive volleys into the white- and blue-coated masses. Twice the French cavalry charged at full gallop and twice they were thrown reeling back. But the Dutch attack on the left had failed completely. Faced by overwhelming numbers, the now sadly depleted British and Hanoverian battalions were forced to retreat, but they did so in good order, facing about every 100 yards or so to discourage pursuit by firing parting volleys.[29] The

Regiment's losses on that grim slope numbered close on 100; its Colonel, by then a major-general and commanding a brigade, had received four wounds.

Unreliable allies and bad generalship had brought about the defeat. But the British infantrymen had even further enhanced their reputation for discipline and courage in the field – their behaviour in camp was at times quite another matter. By today's standards few actions have better deserved a Battle Honour, but again this was denied to the Regiment.

Worse was to come. On 25 July Charles Edward Stuart, grandson of King James II, landed in the Highlands to raise the Scottish clans against the Hanoverian monarchy. Opposed by only the handful of units still remaining in the British Isles, at first he swept all before him. Help was needed, and Cumberland, ordered in September to send ten units home, replied:

> I can assure His Majesty that last Fryday I had the satisfaction to see the whole army under arms, and can with the greatest truth say that the battalions were equally fine and in good order; but if there was any prefference to be given it was to these ten, which I have pick'd out for that very reason.[30]

Among the ten was Colonel Charles Howard's Regiment of Foot.

Cumberland himself, with the balance of the army in Flanders, shortly followed those first ten regiments home, and it was he who bloodily crushed the rebellious Scottish clans at Culloden Moor on 15 April of the following year. The Green Howards, however, took no part. Prior to his defeat the Young Pretender had advanced into England as far as Derby, and the Regiment, with others, had been retained in the Midlands, probably in the hope of intercepting him.

The French took full advantage of the absence not only of the British units, but also a number of German and Dutch battalions which had also been shipped across the Channel to help crush the insurrection. By May they had virtually overrun the Austrian Netherlands, taking Brussels as well as Antwerp. The age-old British fear of an enemy in control of the Channel Coast had been realized. For the time being little help could be given to the Allies, but at the end of June a small force of four cavalry and six English infantry regiments, among them the Green Howards, was spared for the continent, landing in Holland because all the ports further south were denied to it.

After a summer of inconclusive marching and counter-marching, early October found the Allies drawn up to cover Liège, one of the few fortresses still in their hands. Greatly outnumbered and with the River Meuse to their rear, prospects did not look hopeful. In the centre of the Allied line, around the village of Roucoux, were posted four Hanoverian and four British battalions, among them the Green Howards with their Colonel himself there in command. Against this small force the French launched fifty-five battalions. The British and German regiments resisted in a splendid manner, the Green Howards and the future 43rd of Foot being especially commended by their general for 'holding a hollow way until the last'.[31] When the inevitable order to withdraw arrived, the Allied army drew off unpursued and in good order.

This further Allied defeat at Roucoux was followed by yet another in the 1747 campaign. Ten more regiments of British infantry had now arrived, with them Cumberland, now in charge of the Allied armies. Outmanoeuvred, he was brought to battle on 21 June at the village of Lauffeld, just west of Maastrict. Again Howard was in charge of the British infantry which, as before, saw the hardest of the fighting in defence of a huddle of small enclosures divided by high mud walls, topped by thick hedges. In this confused battle, so different from those formal combats of serried ranks more typical of the era, the French lost near double the number of men. In the end the Allies were forced again to retire, a difficult task but once again performed in an exemplary manner. Casualties in the Regiment numbered 165, the highest of any British unit, among the killed being its Lieutenant-Colonel.

The following year brought peace to nations yet again exhausted by war. The end result amounted to little more than the restoration of the *status quo*. With French successes in Flanders offset by British victories in North America, both opponents just surrendered most of their conquered territories.

***

Ordered home in 1748, the Regiment spent six months in the Winchester and Salisbury area, before embarking for its first and short (three years) tour of duty in Gibraltar, for long the key to the western entrance to the Mediterranean, captured from the Spanish in 1704. By then the Regiment had become Lord George Beauclerk's, this nobleman having succeeded Charles Howard to the Colonelcy in 1748.

Because of the survival of its Garrison Orders Book,[32] life on 'The Rock' is well documented. In it can be read many references to the misdeeds of members of the Regiment – some serious others rather less so. Private George Peck, convicted of housebreaking on the evidence of two accomplices who turned King's evidence and were set at liberty, was hanged with due ceremony on the Grand Parade on 7 May 1751. The previous year Corporal William Shaw received 500 lashes and was reduced to the ranks for encouraging two soldiers to fight, while Private Zachariah Rollbred was similarly punished for advising two of his fellows to desert. On the other hand, Ensign William Gunn was merely suspended for a month with loss of pay for making a practice of beating his soldiers, the Governor directing that the forfeited pay be remitted to two of the soldiers who had suffered. Lord George, who had accompanied his Regiment on this posting, took issue that soldiers came to him with complaints 'without first acquainting their Officer', while the Governor threatened with a whipping soldiers who caught young partridges on the hill. The handover between quartermasters when the Regiment was posted back to England in May 1752 sounds much the same as one carried out today, albeit less technical.

In 1751, while they were still in Gibraltar, a Royal Warrant established uniformity of clothing and colours, that of the Regiment being described as green. At the same time, regiments were numbered in order of precedence, Lord George Beauclerk's Regiment becoming officially the 19th of Foot, a name it was to carry for the next 130 years. As with all such changes, the old designation hung on for many years to come.

Back home the Regiment marched from Portsmouth, via Salisbury, to Berwick-on-Tweed, the exact route specified in the War Office Marching Order Book, as was the custom of the day. To North Countrymen it may be of interest that halts were taken at Doncaster, Sherbourn, Tadcaster, Boroughbridge, Northallerton (where a five-day break was granted), Darlington, Durham and Newcastle, not too energetic a pace as Thursdays and Sundays were non-marching days.

From Berwick the Regiment moved again into Scotland where it spent time at both Aberdeen and Inverness, enforcing what were known as 'The Rebellion Acts', keeping order in that then unhappy land and suppressing any lingering embers of the recent insurrection.[33]

★

The short peace after the War of the Austrian Succession was no more than an uneasy interlude in the long-drawn-out conflict between Britain and France. The threat of invasion when the French assembled flat-bottomed boats in their Channel ports in 1775 triggered a belatedly half-hearted strengthening of the weakened armed forces, but the official declaration of war was delayed until the following year.

This start of the Seven Years War, as it was to be termed, saw a further and more worthwhile army expansion, with fifteen line regiments, including the 19th Foot, ordered to recruit second battalions. Raised at Morpeth in August, the 2nd/19th, like its fellows, was transformed two years later into a regiment in its own right, the 66th of Foot, which later became the 2nd Battalion the Royal Berkshire Regiment, now amalgamated into the Duke of Edinburgh's Royal Regiment (Berkshire and Wiltshire). Exactly a century was to pass before The Green Howards would again possess two battalions.

Shipped to the West Indies, the 66th Foot quickly saw some fighting, but its parent spent the first five years of the war on garrison duties, first, as we have seen, in Scotland and then in the north of England and around London. Not that the elder battalion was in any way unfit for action. An inspection at Newcastle held on 21 June 1758 showed a Regiment with 'Yellowish Green' facings and nearly 1,000 strong, 'properly appointed good and well disciplined and fit for service'; its recruits were 'Of a low stature, but young and in General Strong and well made', a not inaccurate description of any latter-day batch of aspirant Green Howards, recruited from the industrial north-east.[34]

During the Seven Years War, the chequerboard of European alliances once again shifted with France linked to Spain, Sweden, Russia and Austria. Prussia, Hanover and Portugal were to fight alongside Britain, but the latter's principal interest lay in acquiring colonies, a strategy that left her in undisputed control of the trading stations of India, many of the sugar islands of the West Indies, and the whole of North America, after Canada and Louisiana had been wrenched from the French. The few British units that fought alongside her German allies on the continent of Europe were of their usual high quality, but the Green Howards were not among them, their single battle being peripheral to the main struggle.

William Pitt, later Lord Chatham, the first and one of the greatest of Britain's prime ministers and an advocate of maritime strategy, who came to power near the start of the war and in the end led the nation to victory, was especially attracted by seaborne expeditions against the French coast. In all, he was responsible for five such operations, the last and the most successful in 1761 against Belle Isle, off the Brittany coast, an island about the size of Guernsey. Part of the 8,000-strong British force told off for the task were the Green Howards.[35]

Well fortified, strongly garrisoned and surrounded by precipitous cliffs, the island was a tough proposition. Two separate landings were made on 8 April, but both were beaten back with heavy casualties, the Green Howards at Port de Andro losing nearly 200 officers and men, including 122 prisoners. A fortnight later another plan was tried, a number of landings at different places with intent to deceive the enemy. Success was gained by a force led by the Regiment's grenadier company, which clambered up a seemingly impregnable rockface, reaching the summit just in time to hold off a counter-attack by 300 Frenchmen until the rest of the Regiment arrived.

The struggle on that cliff top was murderous. A wounded Green Howard subaltern was on the point of being overcome when Private Samuel Johnson rushed to help him, shooting a French grenadier through the head and killing five more with his sword; although wounded himself, he then carried his officer to safety, to be rewarded the next day with promotion to sergeant and the vast sum of twenty guineas, donated by the grateful officer.

Driven into Le Palais, the well defended capital of the island, the French garrison fought until 7 June with great determination until the commander of the citadel, into which the survivors had withdrawn, eventually capitulated. As the French could not reinforce the place in the face of the British fleet, it is hard to see what the object of the operation could have been. The attackers' losses in killed and wounded numbered 739, one-fifth of them Green Howards, the prisoners taken on the first attempted landing being recovered. To the Regiment's grenadier company fell the distinction of taking the formal surrender.

At the war's end Belle Isle was restored to its French owners. More than two hundred years later, in 1951, the Battle Honour 'Belle Isle' was belatedly granted to the seven regiments who captured it, among them the Green Howards.

***

At the end of the Seven Years War the Green Howards were shipped back to Gibraltar where they remained until 1771. Garrison orders suggest that life there was little changed. Lieutenant Huttchison when discovered absent from guard was 'forgiven' by the Governor because of his former good

Above: The Green Howards landing to storm the cliffs at Belle Isle in 1761. A contemporary engraving.

behaviour; a drunken Corporal Dow who subsequently struck an officer was reduced to the ranks and received '1,000 lashes with a cat of nine tails'. Hanging was the penalty for the deserter, Private James Wood.[36]

Garrisoning Scotland and Ireland after its return from 'The Rock', the Regiment almost missed taking part in the American War of Independence. Hostilities between Britain and her thirteen North American colonies had erupted in 1775. With hindsight such a breach can be seen as inevitable. The removal of the French threat during the Seven Years War had led to a rapid deterioration of relations between the four million colonists and a home government, both remote and tactless. It is hardly surprising that attempts to persuade the Americans to contribute towards the burden of debt accumulated by Great Britain during the previous war and to protect the British from the depredations of the native Indian population found little favour.

The difficulties facing the British commanders were immense. Distances were vast and centres of population few. The home politicians burdened them with ambitious plans, based on scant knowledge, but failed to provide the necessary resources; as ever, recruits were hard to come by, resulting once again in the use of large numbers of German mercenaries. Not that Washington and his fellow American generals did not suffer from similar problems of political interference, desertion, disease, poor recruiting and lack of supplies. In the wings the French waited eagerly for signs of American

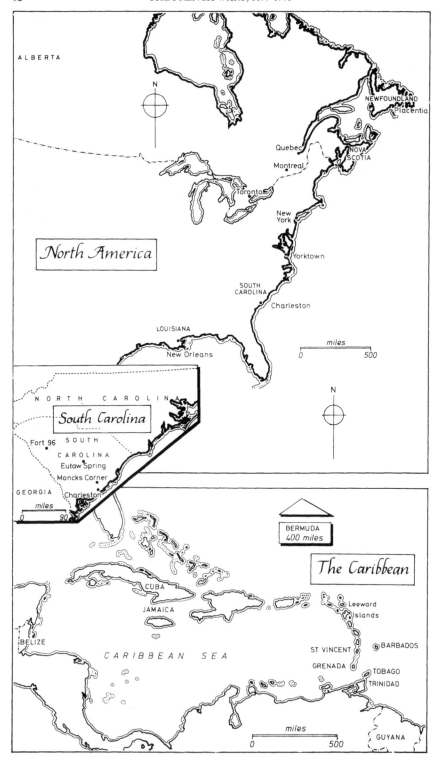

ALBERTA

N

NEWFOUNDLAND
Placentia

Quebec
NOVA
SCOTIA
Montreal

Toronto

New
York

North America

Yorktown

SOUTH
CAROLINA

Charleston

LOUISIANA
New Orleans

miles
0   500

N

N O R T H   C A R O L I N A

South Carolina

Fort 96   S O U T H
C A R O L I N A
Eutaw Spring
Moncks Corner

GEORGIA   Charleston

miles
0   90

BERMUDA
400 miles

The Caribbean

CUBA

JAMAICA

Leeward
Islands

BELIZE

C A R I B B E A N   S E A

ST VINCENT   BARBADOS

GRENADA

TOBAGO
TRINIDAD

miles
0   500

GUYANA

success. Soon after the British disaster at Saratoga 4,000 French troops arrived to assist the Colonists, and the French fleet began to threaten British sea communications across the Atlantic. Subsequently Spain and Holland would join in the coalition against France's enemy.

American Loyalists were numerous, many of them Southerners. With this in mind a British expedition was directed against the Carolinas in 1780, but was to prove a mistaken division of forces. Nevertheless, this southern campaign at first went well, but towards the year's end the British, after a series of savage little battles, were forced back on the defensive. By the following summer the British held nothing outside Charleston except the besieged Fort Ninety-Six, 150 miles up country, garrisoned by 500 Loyalists. The fighting was usually fast-moving, often akin to guerrilla fighting among the woods, warfare in which fieldcraft and markmanship were important, skills the British were learning the hard way under the command of a younger generation of officers who had come to the fore as the war limped slowly on.

The arrival at Charleston on 17 June 1781 of the 19th Foot, together with the Buffs and the 30th of Foot, was opportune, although it was also accidental, an American privateer having intercepted orders to divert them to New York. This reinforcement allowed the very capable British commander, Lord Rawdon, to march immediately with a force some 2,000 strong, including the flank companies of these three fresh units, to relieve Fort Ninety-Six, but the Americans had already raised the siege. The fourteen-day march through the height of the Carolinian summer was gruelling, especially for troops fresh off troopships, fifty men dying of heat stroke. Bringing the garrison with him, Rawdon, at the end of a long line of communications, began to move back towards Charleston.

Holding a staging post at Monck's Corner, about thirty-five miles from Charleston, were the Green Howards, less their flank companies. Lieutenant-Colonel John Coates, in command, mistaking the approach of a raiding party for the entire American army, decided to retreat on Charleston. Burning the church which had been fortified for defence and the stores it contained so as to prevent the enemy using them, the battalion set out on 16 July but was immediately pursued. Captain Bell, writing to a fellow officer, described what followed:

> We thought ourselves unable to cope with the enemy at Monk's Corner, therefore thought it expedient to leave that place, and on the march lost every item of baggage we had. All the officers' baggage, men's tents, knapsacks, blankets, etc., etc., were all burnt, destroyed or pillaged. My trunk unfortunately could not be saved. It, however, fell into General Sumptner's hands [an American], who made a prize of the money, but was civil enough to take care of my books, and Sergt. Laidlaw is gone a few days ago to him to get them again . . . We marched all that night . . . but about nine o'clock a party of the rebels galloped over the bridge in the face of our Field Piece, rode through the Regiment and wounded two men. It was the most daring thing I ever heard of. One of them made a stroke at the colonel, which he turned off with his Hanger. McPhuen instantly brought him down . . . Three of the five paid for their temerity. About this time Colin Campbell with part of the Rear Guard, sick, stragglers. etc., and the wagon with my trunk, were taken

prisoners within three hundred yards of the regiment. They gave them all their parole to go to Charleston.[37]

By disciplined volley fire the Regiment then managed to beat off the enemy. The next morning a relief force arrived from Charleston to extricate them. It had not been too glorious an episode, but the men were, according to American commentators, 'all raw Irishmen' who had seen no service.[38] As well as the sixty prisoners, the Regiment had lost ten men killed and thirty-four wounded; the Paymaster's chest contained 720 guineas, which Sumpter divided among his troops.[39]

The grinding heat now forced both sides to pull back into summer quarters, so allowing some of the many sick to recuperate, but Lord Rawdon himself, broken in health, was obliged to return to England, his command devolving upon Lieutenant-Colonel Stuart of the Buffs. On 8 September the two small armies came face to face with each other at Eutaw Springs, sixty miles from Charleston, upon which Stuart had withdrawn to await a supply column. His 1,800-strong force contained a body of 300 men, the flank companies of the 19th and the 30th, commanded by Major Marjoribanks, an experienced Green Howard officer who had been wounded as a subaltern at Belle Isle.

There is a need to say something about these flank companies – the grenadiers and light company men. The specialist function of the grenadiers had long disappeared, but they were still picked soldiers occupying the place of honour on the right of the battalion's line. A light company had been added to the establishment only in 1771, the product largely of experience in North America during the Seven Years War in which the need for skirmishers had been sorely felt, although such troops had been common in other European armies for the past half century. As with the grenadiers, the better soldiers gravitated towards the light company, its position being on the left of the line. Hence the use of these flank companies, often brigaded, for tasks demanding especial skill or courage.

To return to Eutaw. Stuart had drawn up his force with his right resting on the River Santee where Marjoribank's 'provisional battalion' was echeloned slightly in front of three regular British battalions drawn up in line and astride the main road. In the camp behind was a small body of Loyalist cavalry and some infantry.

When the Americans attacked, their strength slightly greater than that of their opponents, they were driven back in confusion, so encouraging the British on the left to rush gleefully forward. They, in their turn, were thrown back in disorder by the fire of the American second line. Only Marjoribank's men stood firm, repulsing with heavy loss a charge by the American cavalry. Meanwhile the Americans were plundering the British camp and getting drunk on the liquor discovered there, a not too unusual sequel. Attacked again by a force of infantry, Marjoribanks pulled his men back to a stronger position from where he was able to catch the Americans in the flank, capture their guns, and drive them out of the camp. Near the end of the battle he fell, mortally wounded, but his flank companies had saved the day.

It had been a bloody little battle in which about a third of the men on both sides were killed or wounded, but which brought the war in the South to an end, both sides equally crippled. Both claimed it vehemently as a victory. If it had not been for Marjoribanks, Stuart's force would undoubtedly have been defeated.

Among the wounded was another and famous Green Howard, Lord Edward Fitzgerald. Hit in the thigh, he was carried off the field and nursed by his devoted black servant. Later he was to transfer into the 54th Foot where that censorious social critic, William Cobbett, then serving as sergeant-major, described him as 'the only sober and the only honest officer I had known in the Army; [he] was a really honest, conscientous and humane man'. A fervent and courageous Irish patriot, Fitzgerald's end was tragic. Cashiered for fraternizing with Jacobins in Paris, he later joined the United Irishmen, to be mortally wounded by his former comrades in arms while resisting arrest in 1798.[40]

The surrender of the British garrison at Yorktown the following month effectively brought the war in America to an end, but it dragged on elsewhere with the loss of Minorca, then a British colony and important naval base. A compromise peace with America's allies followed in 1783, but the United States of America had been born.

In describing the role of the Green Howards in the final stages of this conflict, the expressions 'British' and 'American' have been used, somewhat inaccurately. British and German deserters were numerous in the ranks of the Americans; large bodies of American Loyalists fought with the British. General Greene, who commanded the Americans at Eutaw, was to write that

Right: Lord Edward Fitzgerald. A gallant officer of the Green Howards and an Irish patriot. (By courtesy of the National Portrait Gallery)

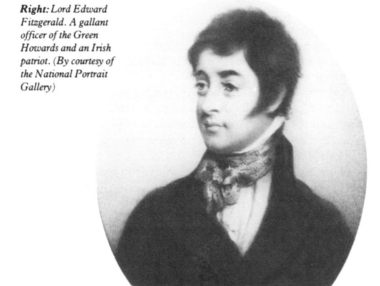

at the close of the war the Americans were fighting with British soldiers and the British with those of America.

Disease finished off most of the survivors of the fighting, the six regiments stationed in Charleston when the British finally quit mustering no more than seventy-five officers and 882 men between them. In December 1782 what little that was left of the Regiment embarked for Jamaica. There they were brought up to strength time and again by drafts. The number of Green Howards who died there from dysentery, yellow fever, ague and malaria is not known, but the Regimental Hospital Books provide some idea of the conditions in that disease-ridden island. Regiments could lose 100 per cent of their strength every two years in the Caribbean, a near death sentence to those posted there and a major deterrent to recruiting.[41]

★★★

The end of the war brought another change of designation, the subject of this War Office circular:

> His Majesty having been pleased to order that the 19th Regiment of Foot which you Command should take the County name of the 19th or 1st York North Riding Regiment and be looked upon as attached to that Division of the County, I am to acquaint you that it is His Majesty's further pleasure that you should in all things conform to that Idea and endeavour by all means in your power to cultivate and improve that connection so as to create a mutual attachment between the County and the Regiment which may at all Times be useful toward recruiting the Regiment.
>
> But as the completing of the several Regiments now generally so deficient is in the present Crisis of the most important national Concern, you will on this Occasion use the utmost possible Exertion for the purpose, by prescribing the greatest diligence to your officers and recruiting parties, and by every suitable attention to the Gentlemen and considerable Inhabitants; and as nothing can so much tend to conciliate their Affections as an Orderly and polite behaviour towards them and an Observance of the strictest discipline in all your Quarters, you will give the most positive Orders on that Head; and you will immediately make such a disposition of your Recruiting parties as may best Answer that end.[42]

Little changes. But so began the Regiment's formal connection with North Yorkshire. It was quite by chance. Prior to introducing the new system soundings had been taken from the Colonel of the Regiment, Lieutenant-General David Graeme, who had taken over from General Beauclerk in 1758. His knowledge of his Regiment's history was scanty, to say the least. Asked whether it had 'any particular Connection or Attachment to Counties; or any reason to wish their Regts to bear the name of a particular County', he replied that it had been raised in Yorkshire and chiefly at Leeds. Nor did he know the date. Nevertheless he seemed pleased at being consulted. 'As Colonel,' he remarked, 'So little is left to my Management, recommendations or discretion, that I have only to bestow my wishes for the Honour and Reputation of the Regmt.'[43] Throughout the 18th Century the powers of a Colonel were gradually being whittled away, and with good reason. Possibly General Graeme was making a point.

3. THE KANDYAN WARS AND NAPOLEON: 1793-1820

Back from Jamaica in April 1791, the Green Howards had returned to a much changed world. The old enemy, differently arrayed, once again threatened Britain's security: three years earlier the French Revolution had created a new regime, one as expansionist as its feudal predecessor.

After the war with America, the British Army had suffered not just its habitual cuts. To an unprecedented extent it had also had been starved of money. Every type of 18th-century abuse had proliferated: incompetents promoted by favouritism, indiscipline, absenteeism, drunkenness and corruption. Military ignorance was widespread. Recruits were near unobtainable and many regiments mere skeletons. With demands for troops ever increasing, the Adjutant-General had plaintively complained 'how we can do it until the Fourteenth and Nineteenth arrive home from Jamaica, I don't know'.[1]

In the manner of such governments, France soon exported its creed and expanded its frontiers. By the middle of 1792 she had declared war on both Austria and Prussia; by the end of that year her new revolutionary armies had repulsed a Prussian incursion and conquered the Austrian Netherlands, the modern Belgium. Feeble and ill-disciplined at first, the French National Guard, a voluntary militia augmented by conscription, was in 1793 amalgamated with what remained of the old royalist army, many of whose officers had been inspired by the revolutionary ideas of the Americans with whom they had fought. Soon they were to conquer most of continental Europe.

Britain's Prime Minister, William Pitt the Younger, son of the Lord Chatham who had led his country to victory in the Seven Years War, warned France that neither the incorporation of the Austrian Netherlands into her territories nor the invasion of Holland would be tolerated. It had no effect. When in January 1793 the French executed their king, pointlessly declared war against Britain, and then invaded Holland, all that Britain could find to reinforce her allies were three battalions of Guards, none of them with transport, medical supplies or reserve ammunition. Their commander was the 25-year-old Frederick, Duke of York, King George III's second son. More troops were scraped together during the summer, including the 19th Foot which embarked on 3 September but was withdrawn from Ostend a few weeks later, having been involved in the capture of the fortress of Menin, an action in which three Green Howards were killed.[2]

The Regiment was to spend an uncomfortable winter. Embarked once again as part of a force under The Earl of Moira (previously Lord Rawdon, under whom it had served in America), it sailed for weeks up and down the stormy English Channel, awaiting the opportunity to land and reinforce the French Royalists in the Vendée whose insurrection was threatening Paris itself. In the end the expedition returned with nothing accomplished, but the unfortunate troops were held in their transports off Southampton, probably to avoid their deserting, where they died by scores, most from typhus fever.

The following June the Green Howards once again landed at Ostend as part of a 10,000-strong force sent to secure the port from a French army threatening to sweep all before it. With them was the 33rd Foot, afterwards the Duke of Wellington's Regiment, then commanded by the young Colonel Arthur Wellesley, in action for the first time. Two regiments with much in common, the Green Howards and the Dukes were to become close friends as the years passed.

Joining forces with the Duke of York's main army, the new arrivals were soon in retreat, their force outnumbered and hampered by the arguments of allies, as ever at loggerheads. At the end of June they split, the Austrians turning eastwards while the Dutch, Germans and British continued northwards into Holland; by October the Duke was holding the line of the River Waal, with a small force in Nijmegen on the south bank of the river and his headquarters in Arnhem.

Seldom had British politicians committed a worse equipped army into action. The many newly raised units lacked effective officers, proper training and even adequate clothing: only the Guards and three line regiments possessed greatcoats, provided by public subscription; the rest made do with flannel waistcoats, bought by their officers. The winter was one of the most bitter on record, the Waal freezing over. Cold and semi-starvation bred typhus and dysentery. At the end of November one-third of the British were down sick. (In the same area exactly 150 years later two other rather better equipped Green Howard battalions were to endure a similar winter.)

In these appalling conditions British soldiers displayed fine courage and determination. When the French crossed the river at Tuyl, twenty miles downstream of Nijmegen, on 27 December, a detachment that included the 19th Foot quickly drove them back. A further crossing at the same place nine days later forced the British to retire, but the pursuing French cavalry were sharply repulsed by a rear guard which included the Regiment and the Black Watch, the action in which the latter won the right to their famous red hackle.

As the army retreated further discipline soon all but collapsed. Troops of the different nations fought one another for food as they scavaged the countryside. An eye-witness described the outcome.

> Far as the eye could reach over the whitened plain were scattered gun-limbers, waggons full of baggage, of stores, or of sick men . . . Beside them lay the horses, dead; around them scores and hundreds of soldiers, dead . . . one and all, stark, frozen dead.

If the retreat had lasted another three or four days it would have matched that of Napoleon's from Moscow. This ignominious campaign ended with the withdrawal of the British survivors from Bremen in April 1795. Exact losses are not known, but are said to have averaged 200 in each unit.[3] It is extraordinary that so many should have survived.

★★★

As this campaign in the Low Countries was ending a young French artillery officer, Napoleon Bonaparte, was starting to carve out the meteoric career which in 1804 would culminate in his being crowned as Emperor of the French – and which, but for Wellington's generalship, Nelson's seamanship and Pitt's diplomatic brilliance, would have left France dominant in Europe and over much of Asia and Africa.

The part played by the Green Howards in these great events was to be no more than peripheral. A year after returning from that grim winter in Holland, they embarked for service in the East Indies, a tropical tour that was to last twenty-four years. In Ceylon (now Sri Lanka) and in India their unpublicized deeds cost them more casualties than fell to regiments that fought through the Peninsula campaign and on to Wellington's final victory over Napoleon at Waterloo. In one way, however, they were fortunate. Four months earlier the Regiment had sailed once again for the West Indies, a far more insalubrious posting, but when severe gales forced the ships back into port, the plans for their future were changed.

For some of the young Green Howards (most of the rank and file were raw recruits), the voyage to the Indies was to prove eventful. The Regiment, less five of its companies, put in at the Cape of Good Hope for water and provisions and there it was detained for several weeks. Both the Cape and Ceylon had been Dutch colonies. Seized by the British after the French had occupied Holland, these two colonies were to be a part of the new Empire. This was based largely on the consolidation of the sugar islands of the West Indies and the expansion of earlier conquests in India – a replacement for the loss of the old empire in North America. The arrival of the 19th at Cape Town in August 1796 coincided with an attempt by the French and the Dutch, now allies, to retake this valuable colony, one that was foiled by the presence of a British fleet. It also provided these young soldiers with both excitement and arduous exercise.[4]

The Green Howards who landed at Colombo in December 1796 were flabbergasted by the island's enchanting beauty. Captain Herbert Beaver, one of several officers who had fought with the Regiment in Holland, enthusing:

> . . . before my room at the Governor's, is the garden – all verdure. At the bottom of it is a sheet of water, varied in its form by luxuriant groves stretching into various parts of it, which renders its irregularities lovely. At about forty miles distant a hilly range arises, sometimes below, sometimes above the clouds . . .[5]

The Dutch, in fact, had held only the coast of Ceylon with its ports of Colombo and Trincomalee, a strip of land in places no more than six miles wide. This 'hilly range' that Beaver saw was part of the wild and mountainous interior, ruled by the King of Kandy, his subjects proud and conceited, taught by past experiences of the Dutch and their Portuguese predecessors to distrust all white intruders.

Although the Regiment was not fated to serve under the future Duke of Wellington in Europe, Green Howards fought with him when he was making his reputation in Asia. In February 1799 five companies were despatched from Ceylon to India to join the forces operating against Tippoo Sahib, the ruler of the southern Indian state of Mysore, one of several supported by French cash and military assistance in their opposition to the British. After a brilliantly organized march, Wellesley, still a colonel, had captured Tippoo's capital, Seringapatam, in a bloody siege in which the ruler was slain, and his French-trained army destroyed. Meanwhile the Green Howards were marching up from the coast, part of a detachment bringing a

Below: The town and lake of Kandy in 1840. Their character and layout had changed little since 1815.

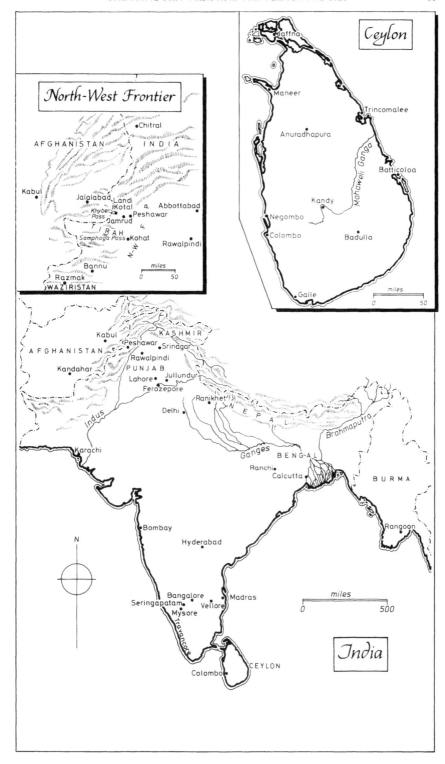

Ceylon

Jaffna
Maneer
Trincomalee
Anuradhapura
Batticoloa
Mahaweli Ganga
Kandy
Negombo
Colombo
Badulla
Galle

miles
0   50

North-West Frontier

AFGHANISTAN   INDIA
Chitral
Kabul
Jalalabad   Landi
Khyber   Kotal   Abbottabad
Pass   Peshawar
Jamrud
Samphaga Pass   Kohat   Rawalpindi
N-W   RAH
Bannu
Razmak
WAZIRISTAN

miles
0   50

Kabul
Peshawar   KASHMIR
Srinagar
AFGHANISTAN
Rawalpindi
Kandahar   PUNJAB
Lahore   Jullundur
Ferozepore
Ranikhet
NEPAL
Delhi
Brahmaputra
Indus
Ganges   BENGAL
Karachi
Ranchi
Calcutta
BURMA
Bombay
Hyderabad
Rangoon
N
Bangalore   Madras
Seringapatam   Vellore
Mysore
Travancore
CEYLON
Colombo

miles
0   500

India

gigantic convoy of 40,000 cattle and 21,000 sheep to feed Wellesley's half-starved force. Earlier they had been used to reduce a number of hill forts, but details of these engagments have not survived. To their chagrin they arrived at Seringapatam nine days after it fell, denied both a share in the victory and the rich plundering that followed.

In August that year those five companies were again in action, detached from Wellesley's army together with thirteen companies of sepoys. Their task was to help quell the Polygars, a collection of robber barons who were ravaging the southernmost tip of India from isolated forts, protected by thick jungle and precipitous ravines. Before returning to Colombo in early 1800 these companies helped to destroy in all forty-four of these forts, capturing, imprisoning and at times executing their Polygar owners. In the heat of the south Indian summer it was gruelling work.

★★★

In 1798 the King of Kandy had died, whereupon his chief minister, or *Mahadigar*, Pilima Talauva, placed an obscure young man upon the throne, so securing the real power for himself. Then began a series of long and tortuous negotiations between the Adigar and the newly arrived British governor, the Honourable Frederick North, later Earl of Guilford, an able and highly intelligent administrator. At first North was not concerned with territorial expansion, aiming only to keep the unstable interior in check by controlling the island's coastal strip and ports. Not that the coast was yet fully under control. In 1800 the Green Howards were used to put down riots in a number of places, during which one of the companies was attacked while marching from Colombo to Negembo fort, one of several of their outstations.

It soon became clear to North that Kandy posed an inherent threat to the island's stability, but he wished to avoid a blatant invasion. Instead, in March 1800 he engineered an invitation from the Kandyans to despatch an embassy to the mountain kingdom, its object to establish a garrison there. Headed by Major-General Macdowall, the mission's escort consisted of fifteen companies drawn equally from the 19th Regiment, a Madras battalion of the East India Company and a locally raised battalion of Malays; in addition there were six guns and a large administrative train.

During a dismal journey up an execrable track into the hills, a succession of deceits practised by the Adigar so aggravated the slow passage of the column that Macdowall was obliged to leave the bulk of his infantry, including most of his Green Howards, in a camp on the edge of the Kandyan dominions. The mission achieved nothing, the Kandyan officials rejecting Macdowall's proposals that a garrison be stationed in their country for the king's protection. But while the talks dragged on, the troops left behind were discovering the perils that lurked in these tropical mountains. Night chills contrasted with the sultry day heat. 'Dysentery, fluxes and liver complaints became frequent; and the jungle fever . . . began to make its appearance.'[6] This 'jungle fever' was, of course, malaria, its cause as yet undiscovered.

The five companies that had suffered with Macdowall were given no respite. A week after their return they were embarked with the rest of the battalion for further service against the still troublesome Polygars.[7] Eight months later, in April 1801, the Regiment was back in Ceylon, landing at the east coast port of Trincomalee and only just missing Colonel Arthur Wellesley who had been there preparing an expedition planned to attack either Java or Mauritius. In the end it sailed for Egypt, but without Wellesley, who for the time being was sent back to India. The 19th was to have been part of this four-battalion force, but their absence in India prevented this. And so it was that the Regiment missed rejoining the mainstream of the war.

***

Although Macdowall's embassy to the Kandyan court had failed, from the military standpoint it had proved invaluable. It had been a full dress rehearsal for the subsequent invasion, during which much had been learned about the terrain, the enemy and the climate – and the diseases as well.

Almost three years were to pass before this invasion was launched. To justify it North used a facile excuse, probably presented to him by Pilima Talauva. The *Adigar* was playing a complicated game. His aim may have been the tempting of the British to destruction in the island's inhospitable interior, the fate of previous Portuguese and Dutch armies. On the other hand he may have been looking for British help to obtain the throne for himself; the king had proved no puppet, but a strong ruler and a danger to the man who had put him there. The tangled skeins of the intrigues have still not been fully unravelled.

The attack on Kandy was double-pronged, one column under Macdowall setting out from Colombo on 31 January 1803 and one from Trincomalee starting four days later. Not counting the usual hordes of followers, a part of any campaign in the Indies, Macdowall's column numbered 1,900 fighting soldiers, including the whole of the 51st Foot, two companies of the 19th, a company of Malays, the Ceylon Native infantry and some Bengal gunners. Another five companies of the 19th, the balance of the Malay Regiment, a single company of Madras Artillery, all under Lieutenant-Colonel Barbut of the 73rd, made up the Trincomalee column. Among the latter's baggage animals were thirty-seven elephants.

Those boyish Green Howard recruits who had landed at the Cape six years earlier were now, to quote their surgeon, 'a remarkably fine body of middle-aged men, who have long been inured to a hot climate, and accustomed to active service in different parts of India'.[8] A similar compliment had been paid during the 1800 riots when two companies crossed the island from Manaar to Trincomalee, the first British troops to do so:

> Though destitute of tents, and with a very small supply of provisions, they persevered, notwithstanding the rainy season had rendered the roads almost impassable. They, however, surmounted every difficulty, and suffered little from Fatigue.[9]

At the same time the Regiment was fortunate to be enjoying an unusual measure of stability among its officers: some fifteen had seen action with it in Holland, while the commanding officer, Lieutenant-Colonel George Dalrymple, was a distinguished fighting soldier, lately of the Black Watch.[10] The British Army was then in the early stages of one of its periodical metamorphoses, a pattern far from unfamiliar. Widespread reforms were being introduced under the pressure of war, largely the result of the vigour and competence of the Duke of York, an inadequate general in the field, but, since 1795, an outstanding reformer as Commander-in-Chief at the Horse Guards. Even as far away as Ceylon his influence may have been felt.

Certainly the arrangements for the expedition were competently executed. Macdowall avoided some of the difficulties of his earlier march by taking a rather longer but less rugged route. There was little opposition. The village people were strangely friendly, and only once did the Kandyans attempt to ambush the long column. Two days later, as the force neared the capital, the grenadier company of the 19th was required to storm a solid stone fort perched on a steep and rocky hill that barred the way, a task it completed in the face of heavy but inaccurate fire at a cost of only two casualties. Barbut's 140-mile march was even less eventful. On 20 February the two columns met, to enter a gutted capital the next day, destroyed and deserted by court and people. Although luck played its part the staff work had been sound. But here was nothing to be looted and a sense of anti-climax prevailed.

The march from the coast had already taken its toll, but now fruitless forays from Kandy against an elusive and now far bolder enemy brought further sickness to already exhausted troops operating in foul conditions. Moreover their clothing, the newly introduced high-collared and tight-fitting coatee and the upright felt shako, were utterly unsuitable for the jungle, even though soldiers did quickly lose their savage leather stocks. Then some of the locally raised troops started to desert, and most of the coolies disappeared as soon as the Kandyans began to ambush the supply columns.

> Parties of banditti hovered continually around the outposts. They concealed themselves in the woods and thickets, fired upon the guards and sentries and whenever any unfortunate sentry fell into their hands, they put them to death . . . One European of the 19th, several Malays, sepoys and coolies, were found mangled in a most cruel manner. The wife of a Malay soldier met with a similar treatment.[11]

The custom of referring to a guerrilla enemy as 'bandits' has a long history.

The Kandyans now threatened the three small forts that had been built on the lines of communication, and the few troops left to garrison Kandy were soon employed in trying to keep the supply lines open. With the military situation fast deteriorating, Macdowall, with North's approval, had already decided to withdraw the larger part of his force to the coast when, to North's surprised delight, Pilima Talauva proposed a truce. Grasping at this opportunity, most of the garrison was thereupon marched out without interference, leaving Barbut in command of a reduced garrison of 300 Green Howards, 700 Malays and a large number of immovable hospital patients. Among those who returned to Colombo were 400 survivors of the 65th; within three months 300 had died from diseases incubated in the hills.

Right: General the Honourable Sir Charles Howard, KB. Oil painting by Wooton and Hudson. The troopers are in the uniform of the 3rd Dragoons, of which he was also Colonel. The background in Carlisle Castle.

Below: 'The Dinner Round' by C. B. Newhouse, showing various uniforms of the Light Company in 1836.

Soon Barbut was evacuated to the coast where he died a few weeks later. Macdowall then returned to Kandy to take Barbut's place and be close at hand for further negotiations with the Adigar. What he found there was described by a Green Howard officer:

> . . . our mortality and sickness is every day increasing, such is the meloncholy state of our detachment, that out of two hundred and thirty-four men remaining of those you left behind, there are not above five fit for duty, and even their services are required to attend those who are in hospital. The number at present in the hospital is one hundred and twelve, mostly fevers, and fifty sick in barracks. The detachment at Fort Macdowall, have only eight men, out of fifty, fit for duty. Yesterday, on the arrival of Gen Macdowall, he ordered the whole of the convalescent sick to get ready to march for Trincomalee; but, dreadful to relate, when they came to be mustered, only twenty-three men out of the whole were found to be able to march.[12]

With little news from Pilima Talauva, the truce was collapsing and the Kandyans were again harrying the tenuous supply lines. Macdowall, himself very ill, returned again to Colombo to persuade North that somehow the survivors must be evacuated, leaving in command Major Adam Davie of the Malay Regiment, an officer with experience neither of warfare nor the country, remembered by fellow officers merely as a well-disposed, inoffensive sort of person. Anticipating what might well happen, Davie had asked if he might decline the appointment, his grounds being that it would bring him only discredit and blame. In his last personal letter from Kandy he wrote:

> Henderson died on the 11th, and Bausset this morning: Rumley and Gonpil are also ill. The Lascars and Malays desert by dozens, and high rewards are offered to murder all the officers. – Batteries close in on us. – Our bullocks carried off by force . . .[13]

The end came on 24 June. Before dawn a vast horde of Kandyans rushed the rudimentary defences among the ruins of the palace, which only twenty Europeans retained sufficient strength to help man. Nevertheless a few Green Howards, aided by a single round of grapeshot, drove out the attackers, but only after one officer had been stabbed to death by a kris. The Kandyans did not try again but instead poured a hail of fire into the perimeter from their own guns and a captured battery.

At about midday, Davie, encouraged by four British fellow officers of the Malay Regiment, went out to the Kandyan lines carrying a flag of truce, but without, it would seem, having consulted the surviving officers of the 19th or of the 51st Foot. Terms were quickly agreed between Davie and Pilima Talauva, now commanding the Kandyan army. Those troops capable of so doing were to be allowed to march unmolested down to Trincomalee. The sick would be cared for in hospital until arrangements were made to evacuate them.

Top left: 'The XIXth Foot waiting at Guy's Hospital in November 1852 to fall in for the funeral of the Duke of Wellington'. From a Sketch by R. Ebsworth.

Left: Light Company, 19th Foot, forming a flank for the Regiment. Light Division manoeuvres, Chobham Camp, 1853. Watercolour by A. F. de Prades.

Late that afternoon, thirty-four Europeans, 250 Malays, and a small party of Bengal gunners began to trudge down the two-mile-long main street of the city in deluging rain; with them were some Malay and Indian wives, with a few children. That night the party reached the banks of the Mahaweli Ganga, only to find the river in flood and neither boats nor rafts available. There they waited, closely watched by the Kandyans. The following afternoon one of the sick Europeans who had been left behind in hospital, dragged himself into the camp with a horrifying tale.

No sooner had the garrison left the town than a vast body of Kandyans had rushed into the palace and butchered in a revolting fashion all the helpless soldiers lying in the hospital. The number slain was uncertain, but probably numbered between one and two hundred. Only two men survived, the one who made his way down to the river and one other, discovered the next day by the Kandyans under a pile of dead.

Throughout that day by the river bank, the soldiers had laboured to build rafts; for a time a few Kandyans had helped them but they soon disappeared, promising to bring boats the following day. When the troops awoke next morning, shivering in the dampness of the dawn, they found, not boats, but a vast and menacing crowd of Kandyans surrounding them. Nevertheless, some of the Bengal gunners managed to swim a warp across the river, only to see it severed by some of their enemy on the other bank, a blow to morale which started a steady trickle of desertions among the Malays. The Sepoys, however, stood firm. What happened next is not altogether clear, but emissaries seem to have delivered an ultimatum to Davie to surrender his weapons and march back to Kandy. This resulted in his calling his officers to a council of war, after which orders were issued to lay down arms, a request by two old soldiers of the 19th Foot to hold their separate council being refused.

The Kandyans then separated the Asian troops from the British, and a body of Africans – termed Caffres – in Kandyan employ then began to lead the white soldiers in pairs towards a hidden hollow near the river bank. Once there the Caffres chopped them down with large swords. Meanwhile the officers were being slain separately, although three Green Howards managed to commit suicide. As Davie and another officer, the last left alive, were about to be murdered, Pilima Talauva arrived on the spot and ordered them instead to be sent to the king. Two other officers managed to hide but were found four days later: one died in captivity, but the other, the 19th's surgeon, escaped to the coast a year later. Davie was spared and lived on in a loose form of captivity for ten wretched years before he died.

One soldier survived. In that gruesome hollow, Corporal Barnsley of the 19th received a blow to the back of the neck that failed to kill him. Waking when it was nearly dark, he found himself stripped to his shirt and able to walk only by supporting his head in his hand, a tendon in his neck having been severed. The next morning Barnsley made his way down to the river, the level of which had dropped during the night, and just managed to swim across. He then struggled on in agony for thirteen terrible miles towards Fort Macdowall. Intercepted by a large body of Kandyan troops, he was treated

well, given some food, and sent off to the fort with the extraordinary message
that its defenders should come out and fight the Kandyans in the open. An
eye-witness described his arrival:

> The sentinel was struck with terror at the emaciated figure and ghastly look;
> he was conducted to Captain Madge, commander of the Fortress at the time,
> who was thunderstruck by his appearance, and the melancholy tidings he bore.
> The first words he said was 'The troops in Candy are all dished your honour'.[14]

Stragglers habitually exaggerate ill tidings. This one did not.

Half-heartedly besieged for the past three days, the garrison numbered
twenty-two Malays and thirty-two Green Howards, nineteen incapable of
moving and the rest all sickly or convalescent. That night Madge evacuated
the fort, abandoning his invalids, although Barnsley had told him the fate of
the patients left behind in Kandy. For four days the Kandyans harassed
Madge's withdrawal, conducted in a masterly manner for which he received
well-merited praise. On the other hand he received no censure for leaving
the sick to their fate. Of the many writers who described the events, only
one hinted at a reproach.[15]

It is nice to know that Barnsley survived. Promoted sergeant, he was
broken for drunkeness. Invalided home in 1805, he ended his days as a
member of a veteran corps in Fort George, Inverness.

★★★

Not only had the British been expelled ignominiously from the Kandyan
kingdom, but few troops were left to repel the seemingly inevitable invasion
of the coastal area. At the start of the year the British garrison had numbered
5,000 men. By the end of June 2,000 were either dead or prisoners. Of the
survivors, most were sick and many more would die before the year was out.

Fortunately for the British, the Kandyans held their hand while they
celebrated their festival of *perahera*. As soon as this was over, however, they
swarmed down from the hills and across the coastal provinces, to surge
around the walls of the small forts, whose garrisons Macdowall had
strengthened during that very welcome lull. But the British did not sit behind
their fortifications waiting to be attacked. Small parties of Sepoys and
Malays, stiffened whenever possible by a few survivors from the British
regiments, were scraped together to make vigorous and successful sallies
against the Kandyans. Especially prominent in this aggressive patrolling was
Captain Herbert Beaver who had distinguished himself in the earlier fighting.
He now won further renown. His descriptions of this jungle and paddy field
warfare are vivid:

> The Cingalese lie concealed till you come close upon them, then they give one
> regular fire, and fly; this is the general case, and I suppose I was about six
> yards from their grasshoppers [or gingals, man-portable pieces of light
> artillery], the balls of which are about an inch in diameter, when they let them
> off. We were attacked from three points at once, but immediately carried them
> all . . . they pick off at least one European or two, from me, at each encounter,
> and we are obliged to be in advance, and consequently it is only now and then

that a Sepoy or Malay is killed . . . I cannot give you an idea of the country; the jungle is so thick and the fastnesses so strong, that we are not for a moment sure but what we may be destroyed by a masked battery. My whole force now consists of only 60 Europeans, 140 Sepoys and 170 Malays.[16]

In this way the British did more than hold their own. By October they had pushed the Kandyans back into their own country. It was only then that reinforcements began to arrive from India, including 300 men of the 10th Foot drafted to the 19th and 51st. General Arthur Wellesley (as he now was), preoccupied with operations against the Mahrattas, was far from pleased at what he saw as North's disgraceful folly in embarking upon an unnecessary war.

Heartened by these successes and the arrival of fresh units, but unaware of the severe reprimand making its slow way to him from the Colonial Secretary, early in 1804 North planned a further offensive against Kandy. In order to ease slightly the problem of collecting enough of the essential porters, this time a number of converging columns were to move simultaneously from seven different points on the coast. One of these columns, starting from the pleasant little west coast port and harbour of Batticoloa, was commanded by a Green Howard, Captain Arthur Johnston, a courageous, able and intelligent young officer. Unfortunately the two written instructions he received were both slovenly and imprecise.

Johnston's force numbered 300, one-third of them Green Howards and the rest Sepoys and Malays, plus a small Royal Artillery detachment. Only 550 porters could be collected, half the number needed. On 20 September they set out. As they entered the Kandyan kingdom they traversed a wild and hilly region utterly devoid of inhabitants, some of it even today unmarked by roads. After marching for 180 exhausting miles in those wretched uniforms through forests and across mountain ranges, burdened by eighty rounds of ball and three days' rations, they arrived in the Kandyan capital on 6 October. During their journey they had fought a number of ferocious little actions, including a river crossing during which two sturdy Green Howards, Simon Gleason and Patrick Quin, swam the Mahaweli Ganga under fire to snatch a boat from the far bank, a feat that could have brought them the Victoria Cross if such a decoration had then existed.

The night before the British entered the city they were heartened by the sound of gun fire. Other columns, they assumed, had also arrived. They quickly discovered their error. The town of Kandy was deserted except for one old woman and one small boy. The next evening an escaped Malay prisoner brought disquieting news: the other columns had all been repulsed. Johnston's position was unenviable. His ammunition and supplies were running low, he was encumbered with sick and wounded, his troops badly needed rest, and the north-eastern monsoon was setting in early. Moreover, morale was faltering. The troops were occupying the same buildings in which their comrades had been butchered: 'They saw displayed in savage triumph in several of the apartments of the Palace, the hats, shoes, canteens and accoutrements of their murdered comrades, most of them still marked with the names of their ill-fated owners'.[17]

After only fifty-six hours Johnston left the capital but made for Trincomalee, a rather shorter journey than returning to Batticoloa. Reaching the Mahawali Ganga, which they were to raft across, they came across the skeletons of Davie's slaughtered officers hanging from trees, and the bones of his soldiers littering the ground. For ten terrible days the column fought its way down the 140-mile track to the coast, harassed by a nimble enemy made bold by success. Sweltering in the forest by day and soaked to the skin by night, subsisting on a little raw salt beef and mouldy uncooked rice, cases of dysentery and malaria multiplied. Struggling under the burden of their sick and wounded, the men became progressively weaker. With a horrible death awaiting anyone captured, the wounded Lieutenant Vincent was carried with an open knife in his hand, ready to kill himself if need be. It may have been premonition. He was among those who were taken by the Kandyans in the confused fighting.

So weak from dysentery that he was obliged to let his men carry him in his cloak, Johnston had been tortured by the fear that his decision to quit Kandy had been wrong. His mind was put at rest when the column at last struggled in to Trincomalee. That second ineptly muddled instruction he had received was intended to cancel the operation.

It was a tribute to Johnston's fine leadership that he lost in all only thirty-eight officers and soldiers killed and missing, nine of them Green Howards. But, as so often happened, of those who reached the coast, 'almost all died in the hospital, few, very few, survived'.[18] Johnston lived, to become commandant of the Senior Wing of the newly instituted Royal Military Academy, the embryo Staff College. His suitability for the post is apparent from the proposals he made in the book he wrote about the expedition. He criticized the clothing, the weapons, the transport system and the European officers' ignorance of local languages, but perhaps his most perspicacious comment was that as each British soldier had cost the taxpayer at least £100, the objects of the campaign might have been better achieved by bribing the right people at the Kandyan court.

★★★

During the following year the war faltered to an indecisive conclusion with the British carrying out wide-ranging raids. In these they devastated the Kandyan countryside, but accomplished little else, unable to compel their resolute opponents to surrender their country. A new Governor, Major-General Sir Thomas Maitland, a shrewd and tough Lowlander, was to put a stop to the war and at the same time cut the overblown and far from efficient military establishment, collecting the regiments from their isolated posts and restoring their discipline.

There is small doubt that the Green Howards were, by then, also in a sorry state. A newly arrived ensign, Edmund Lockyer (who was in later life to claim Western Australia for the British Crown) described his reception in Ceylon just after Johnston returned. Met at the quayside by a Captain Alexander Lawrence, an elderly officer shabbily dressed in blue coat and trousers, Lockyer was conducted to Lieutenant-Colonel William Vincent, a pleasantly inoffensive and half-blind old gentleman whose unit was run by his quartermaster and his adjutant.

The former, Lieutenant John Crooks, had an unfortunate touch. He was to greet Johnstone's handful of survivors with the words 'What a nice, dirty set of fellows, you are a disgrace to any Regiment', a remark that provoked one soldier to retort 'By God, Johnny, had thee been where we have been, you would not now be here.' The court-martial was inevitable. Gossip had it that this Crooks, promoted from the ranks, had obtained his commission by bribing Colonel Vincent. Captain Lawrence was a braggart with sabre cuts on both cheeks and a couple of missing fingers, wounds received in years of hard campaigning, his wife 'a gaunt, tall, vulgar Irishwoman, with half-a-dozen plain, sickly-looking children'. One of those children became Sir Henry Lawrence, Bart, killed in the Indian Mutiny at Lucknow; another was to be Lord Lawrence, the illustrious 'Lion of the Punjab'; the eldest merely achieved the rank of colonel, a fourth received a knighthood and another ended his career as a lieutenant-general.[19]

To return to Maitland. So well did he succeed that in August 1806 the Green Howards were capable of leaving Ceylon at only two hours' notice for Vellore in southern India. A major and bloody mutiny had erupted among the Sepoys of the Madras Army, as a result of ill-conceived reforms instituted by newly arrived officers, but it had been quelled by the time the 19th arrived.

The Regiment was to remain in India for a year before returning to Colombo, but in 1809 it once again crossed the sea back to India, this time to take part in the subjugation of the Rajah of Travancore who, after neglecting to pay his subsidy to the Madras Presidency, plotted to murder his British Resident. After a setback on the evening of the Battalion's arrival, in which several men were killed and wounded, it helped capture, in what was described as 'gallant style', an enemy stockade and extensive breast-works. Soon after this action another major mutiny broke out in the Madras Army, largely the consequence of the tactless handling of the 'John Company' officers by, among others, that self-same General Macdowall, lately of Ceylon.

Lieutenant-Colonel the Honourable P. Stuart of the 19th (afterwards a full general and Governor of Malta) suffered in the subsequent operations from having his column threatened by the Travancore insurgents on one side and the mutineers on the other, but in the end he received credit for helping to conciliate the mutinous 'John Company' units. As so often happens in such small wars, officers, both senior and junior, were required to show diplomatic as well as military skills.

***

After so long in the tropics the Regiment had in many ways adapted itself to the life. By 1814 private soldiers were enjoying fish or beef curry three times a day, accompanied by coffee and rice cakes for breakfast and supper, and vegetables for dinner. The NCOs were served with eggs and fruit as well, all a far cry from the daily pound of bread and three-quarters of a pound of meat, the immutable ration of the British soldier elsewhere until the middle of the century (there is a need to remember how welcome even that was to many half-starved English farm labourers who rarely tasted meat, or Irish peasantry subsisting on potatoes and buttermilk in their turf cabins). A dram of arrack was also issued twice daily, diluted in the case of the young drummers, recruits and 'notorious drunkards'. Clothing also was on a generous scale, each man having five white and two fatigue shirts, five pairs of nankeen and two fatigue trousers and three white jackets, a mark of how concessions were at last being made to the climate. Gingham or dungaree trousers had also been tried but the material had fallen into holes.[20]

The Diary of Colour Sergeant Calladine describes life in the Regiment, both in Ceylon and elsewhere. An intelligent and reasonably educated apprentice to a framework knitter, Calladine joined the army even though he was earning good money. Throughout his twenty-seven years' service he kept this detailed diary, one of the few first-hand descriptions of life in the ranks at the time. Why he took the King's shilling, he never properly explained: perhaps it was adventure, there are hints of girl trouble, possibly it was a combination of the two. Enlisting in the Militia in 1810, he transferred to the 19th two years later on the new seven-year short service engagement. For two years he served with the Depot, for a time at Hull, where most of his comrades were volunteers from either the Tower Hamlets or North Yorks Militia. At last, in May 1814, he embarked in a large draft for Ceylon. A spendthrift young man, his conduct was at times far from perfect, probably the reason why his first stripe did not arrive for eight years.

It is clear from Calladine's journal that relations between the officers and their soldiers were by now quite close, not only in Ceylon but also when the Regiment was scattered in small detachments on quasi-police duties in Ireland and the North of England. Sport of a sort had begun: when stationed at Chelmsford, a Derby Militia team beat the town and county at cricket. Calladine sang in the detachment choir at Batticoloa, a pleasant station where '. . . all kinds of provisions very reasonable, duty easy, liquor cheap and good, and we had all the indulgences that soldiers could expect'.[21] But drink, as ever, was the main amusement, as it was among all ranks of society, a besetting national sin.

Death from tropical diseases, aggravated by drink, was, of course, an ever-present threat. During the Regiment's twenty-four year stay in Ceylon, 1,498 men died either by disease or in action, an annual average of seventy-six in each thousand, with twenty-seven more invalided home, a total wastage of 10 per cent each year.[22] Nevertheless, a high proportion of the losses occured either during or immediately after the various jungle operations. If those years are ignored, the annual ratio of deaths comes down to thirty-seven in each thousand, a very different figure, one that can be compared with the death rate in England, about one in forty in 1810.

★★★

Wellington's complaint that his soldiers were 'the scum of the earth enlisted for drink' is celebrated, less so his qualification that 'it really is wonderful that we should have made them into the fine fellows they are'. But a significant number was of a quite different type; not just society's outcasts and half-starved Roman Catholic Irish peasantry (since 1780 again permitted to enlist), but, like Calladine, volunteers from the Militia, men tempted possibly by the prospects of adventure, more often by the cash bounty, but sometimes by the wish to defend their country against Napoleon. More than anything the short service seven-year engagement, introduced in 1806, persuaded such men to transfer; before this recruits had enlisted for life, as they were to again when short service enlistment was abolished.

Discipline was as harsh as ever, but the officer who conducted the 19th's 1813 annual inspection noted that corporal punishment only made the depraved and hard even more so. He recommended 'more humiliating punishments'. Solitary confinement with drills could be more efficacious, together with the chaining of an offender's leg to a shell or log. He commented that account keeping had taken precedence over general discipline. The officers, he remarked, 'mess together and conduct themselves with the greatest propriety', but the soldiers had heavy debts.[23]

★★★

A few words should be said about this Militia. (Mention has been already been made of Luttrell's and Erle's Militia units, and of the men from the North York Militia who had transferred to the 19th Foot.) After the outbreak of the Seven Years War in 1757, William Pitt had introduced his Militia Act in an attempt to convert the Militia to a less inadequate force. This Act was based upon the ancient principle that every man had a duty to defend his country, something understood upon the continent of Europe but less appreciated in Britain. Under this new Act the county rates would pay for this force, the quota to be raised in each county would be defined, and men were to be selected by ballot, although the purchase of substitutes was permitted and was to become commonplace. Officers would qualify for commissions by holding or being heir to property, so placing control firmly in the hands of the landed gentry. Dryden's criticism of the old Militia would often still be apposite:

> Mouths without hands, maintained at vast expense In peace a charge, in war a weak defence.

Nevertheless, when invasion threatened, volunteers were rarely lacking. Although the Militia was not required to serve abroad, it was the prime force available for home defence in the absence of the regular army overseas or any proper police force.

Embodied in 1759, the North York Militia at first had two battalions, the Richmondshire and the Cleveland and Bulmer. It was soon in action but against its own countrymen, the Northumberland pitmen who were rioting in protest against the self-same Militia call-up. Twenty-one civilians perished on the first occasion, and about fifty on the second, two members of the Cleveland Battalion, including an ensign, being killed and four wounded. This ended the riots but gained the unit the name of the 'Hexham Butchers'.

Called out again during the War of American Independence, it is likely that it helped to crush the famous London 'Gordon Riots' of June 1780. Then, back in the north, it supressed severe violence in Sunderland, directed against pressing local men for the Royal Navy. With most of the regular army absent overseas, more often than not the Militia had to handle this unenviable 'peace keeping' work.

Inspection reports at this time often describe these Militiamen as tall, well-limbed and stout, an indication that in the North Riding living standards were sounder than elsewhere in a country where rural poverty was damaging health and physique.[24]

Called up again for the French Revolutionary and Napoleonic Wars, the North York Militia alternated between coastal defence, guarding prisoners of war, suppressing riots and even fire-fighting, from Kent as far as Scotland. Increasingly its role became that of feeding trained men into the regular army. In 1809 alone, 280 men transferred in this way.

It is little known that in 1795 the two newly raised light companies of the North Yorks Militia adopted green uniforms, the first British troops to wear that famous colour. When reviewing the battalion the Duke of York took especial notice of the innovation. Tradition has it that one of his ADCs, who later formed an experimental corps of riflemen, the forerunner of the Rifle Brigade and the Royal Green Jackets, after seeing this parade decided to dress his men in green.[25]

To meet the very real threat of a French invasion, units of Volunteers were also raised. In the North Riding the first such unit seems to have been The Scarborough Volunteer Infantry, formed in 1794 and consisting of three companies. The healthy sum of £1,129.5s.6d. was raised locally by public subscription to dress and maintain them for two years; arms, ammunition and accoutrements were to be provided by the government, an obligation upon which it defaulted, finding no more than eighty stands. Although designated infantry, they also handled the guns mounted on the Castle Hill. Colours were presented by the local ladies, a local innkeeper is recorded as having played the bassoon in the band, while their stirring song, an adaptation of 'The British Grenadiers' ended:

> Should France persist we'll ne'er desist
> To thunder in her ears,
> With full three hundred gallant lads,
> All Scarborough Volunteers.[26]

Other North Riding units, the forerunners of today's Territorials, were soon learning the elements of their drill, among them the Loyal Dales Volunteers, the Catterick and Richmond Regiment and the Masham Independent Company, the latter manned by solid citizens who could provide their own uniforms and accoutrements. Fear of invasion, civic pride and delight in military swagger and uniform all played a part in encouraging men to enlist, although some did so to avoid service in the Militia. There was a problem with many of the Dalesmen. Intermarriage was such that surnames were few. A muster call revealed twenty-two Thomas Aldersons and fifteen John Hirds.

Recourse was made to the formal use of the nicknames by which they were known in their villages, so that the parade ground at Richmond Castle echoed to calls for 'Tripy Tom', 'Black Jack', 'Slipe', 'Split Meat' and 'Shodder'.[27]

Units were embodied for a period in 1804, and on one occasion they were marched off towards their battle positions on the coast, only to be turned back at Thirsk when it was learned that the warning beacons had been lit in error. But, even before Nelson's destruction of French sea power at the Battle of Trafalgar in 1805, Napoleon had abandoned his plans to invade England, emptying the invasion camps in which 200,000 men had collected, fearsome opponents indeed for these enthusiastic but scarcely trained Militia and Volunteer units. Soon, with invasion fears removed, the Volunteers disappeared, to be revived once again half a century later upon renewed fears of French aggression.

***

In Ceylon the Kandyans remained a menace – one which Britain was chary of tackling, fully stretched as it was in a major war. In 1812 Sir Thomas Maitland, who had avoided conflict with Kandy while placing the Colony's administration upon a sound footing, was succeeded by an able but rather elderly soldier, Lieutenant-General Sir Thomas Brownrigg, lately Quarter-master-General to the Duke of York, a post then analogous to Chief of Staff. Brownrigg's instructions from Lord Liverpool, the Secretary of State for War and the Colonies, were in no way ambiguous:

> If War should, after every endeavour on your part to avoid it, be rendered unavoidable by the Acts of the Candian Government, the mode of conducting it must of course be left entirely to your own discretion; but I trust you will agree with me in the Opinion, that the same Principles which have induced the British Government in Ceylon to deprecate the commencement of this Contest should continue to operate in the Conduct of it; and that every Measure of offensive hostility that may be resorted to should be undertaken solely with the view of providing for the Security of our present Possessions, and not for the Extension of them; and that, on the attaining of that Object, it will be most desirable to attempt the renewal of our former good understanding with the Candian Government with the least delay.[28]

Despite this warning, for such it undoubtedly was, Brownrigg was determined to complete the work which North had tackled so ineffectively a decade earlier. The temptation for the new Governor to interfere was strong. Kandy was in chaos. Pilima Talauva was dead, executed by that puppet youth of a king whom he had placed upon the throne. Cruel and ruthless, the monarch had subsequently alienated many of his chiefs and headmen, some provinces were in revolt, and several of the country's leaders, including Pilima Talauva's successor as First Adigar, were in touch with the British. Then, in October 1814, the king thrust back across his border ten native traders from British Ceylon, their severed noses, right ears and right arms festooned about their persons. It was the opportunity for which Brownrigg had waited. His invasion plans completed, in January of the following year he launched eight separate columns into Kandy.

The plan was very similar to the one aborted in 1804, but this time it was far better organized. Brownrigg had read Johnston's book and was receptive to new ideas. As well as the 19th Regiment, the garrison contained one other British battalion, the 73rd Highlanders, later the 2nd Black Watch; the Ceylon Regiment, now four battalions strong, was manned by Malays, Sepoys, black African slaves and Javanese. The eight columns were mixed, part native troops, part European, with either Green Howards or Black Watch in each.

The march involved the usual hardships, accentuated by the outbreak of the monsoon and the unfitness of some of the troops and the pack-bullocks. For Private John Kirner of the 19th Foot it was especially tough: he endured it all with a flayed back, having received six dozen lashes for drunkenness on the second day out. Nevertheless in less than a month the first of the converging columns entered the Kandyan capital (deserted as ever), having met no more than derisory opposition. Provinces had surrendered as the troops advanced, and individual Kandyans soon flocked back to the city. On 18 February the king was captured, after patrols had hacked their way through the mountainous jungle, men often covered in blood from head to foot, the effect not of enemy bullets but the ubiquitous leech.

Not a single soldier had been killed, and deaths from sickness were unexpectedly few, the dreaded jungle fever having failed to appear. Brownrigg had good cause to congratulate himself. Kandy was then annexed to the British Crown and the king in due course departed for exile in Vellore. His captors had respected the pride and fortitude he displayed, as well as the nice sense of humour shown by a monarch who was better known for the pleasure he found in impaling his enemies.

A garrison of 800 men was established in Kandy, with smaller ones elsewhere. Strangely enough Kandy proved to be a very healthy station, possibly because, in improving his capital, the king had drained the swamps and so killed the malaria carrying mosquito.

The future Colour-Sergeant Calladine, whose draft had arrived during these operations, enjoyed both the double ration of arrack issued to celebrate the king's capture, and watching his arrival in Colombo with his retinue. Brought round by sea in due course to join his unit at Trincomalee, Calladine then entered hospital with dysentery, as did many of his draft, joining there a large number of men who had returned from the interior, weakened by their exertions and carrying the dormant germs of disease. In contrast to the comparative good health they had enjoyed up country, each day four or five of Calladine's fellow patients were carried out for burial.

★★★

British rule in Kandy was mild, if not indulgent, and relations between the Kandyan ruling classes and the newly imported European officials were superficially sound. But the Sinhalese were a proud people who had successfully repelled attempts first by the Portuguese, then by the Dutch and latterly by the British to conquer their country. All seemed tranquil, but the underlying discontent was widespread and bitter. When Mr Wilson, the British Agent at Badulla, was murdered in October 1817 the revolt rapidly spread.

For the 19th Foot, at long last under orders for home, the timing was inopportune. The women of Calladine's detachment had already left to join the Headquarters, and he remembered:

> . . . what high spirits we were all in, thinking soon to follow them to Trincomalee, and from there to embark for old England; but alas! very few, if any, of these women ever saw their husbands again . . .[29]

Commanding the garrison at Badulla was Major Macdonald of the 19th, a most capable officer. Marching his small force to avenge Wilson, in two days he covered twenty-four miles *as the crow flies*, across the grain of the most rugged mountains of the islands, fighting a number of small actions on the way in which he himself was twice hit by arrows. All he could do was to lay waste the countryside, but this he accomplished to such good effect that several headmen surrendered. Many British officers assumed that this rapid swoop by Macdonald had crushed the rebellion. Brownrigg was less sanguine. Every available soldier was ordered up country, including Calladine's company from Batticoloa.

Calladine has left us a graphic description of the march, which followed much the same route as Johnston's, thirteen years before. So heavily laden were the men that after a short time they threw away half their kit, so that 'the place was strewn with shirts, trousers and shoes'. Devoured by mosquitoes, they saw little of their opponents except when sniped by arrows. For much of the way they marched barefoot to avoid pulling off their shoes and gaiters at every stream they crossed; to do otherwise was to be quickly lamed by the sand and stones that found their way inside. Like Macdonald's party, whenever they passed through an inhabited area they burned every village they came across.

Terror bred terror. Before long, atrocities multiplied on both sides, to the disgust of many officers, though not of Captain Ajax Anderson, a versifier of the 19th Foot, who declaimed:

> We'll track the savage to his den,
> With famine, sword and flame.[30]

A private soldier of the 19th Foot saw it otherwise. 'When I am discharged', an officer heard him remark, 'I intend to become a highwayman; for one thing, after what I have seen in Kandy, taking the life of a man will give me no concern.'[31]

The British despatched patrols in all directions, but their progress was hindered by the difficulties of carrying their wounded, and the need to stop and burn the body of any soldier who was killed so as to prevent the Kandyans impaling the head on a post close to a British camp. At first Brownrigg was surprised how healthy his soldiers remained in such conditions, but soon disease began to take its usual toll. A fifty-six hour patrol from Kandy by a detachment of 100 Green Howards resulted in every man being admitted to hospital within the week.

Soon it became dificult to maintain outlying posts, and by March many had been abandoned. Casualties by arrow and bullet mounted as the Kandyans became bolder. With the British troops tiring, the Kandyans massed their forces and half-heartedly attacked those posts still garrisoned. All failed. Kandyan losses were not serious, but their morale was probably critically damaged when they withdrew with nothing accomplished.

In one such action Major Macdonald with only eighty men was beseiged by 7,000 or 8,000 Kandyans, some half of them armed with muskets. From a weak position on a low-lying hill, these few Green Howards held off the enemy for ten days by bold and regular forays made by small parties of men. Utterly exhausted though the defenders were when the Kandyans at length withdrew, they had not suffered a single casualty.

By May 1818 the Kandyans were in complete control of large stretches of their country. With the hospital in Kandy holding 500 patients, morale there was low indeed; the débâcle of 1803 was in everyone's mind. So it was fortunate that the Kandyans, suffering terribly in every way, were losing heart. With Napoleon beaten, troops could be spared from elsewhere, and soon fresh battalions were arriving from India. The British were winning a war of attrition. Then the Kandyans committed a major error. They abandoned their successful guerrilla tactics and instead concentrated an offensive force, 2,000 strong; this a small British column quickly routed and then pursued for ten days. Units had now learned to live in the jungle and move light, carrying only their haversacks.

By September the rebellion was nearly over. This Third Kandyan War, as it is sometimes described, had been crushed at a cost of no more than forty-four officers and soldiers killed by bullets or arrows. One in five of the Europeans had, however, died of disease during 1818 alone, many of them men newly arrived from Europe and who had survived Waterloo. Of the Green Howards, five officers and 114 men had perished. Calladine was fortunate, surviving several attacks of fever, ague and beri-beri, the last named cured, so he persuaded himself, by regular doses of arrack. Back in Colombo he convalesced, like many others receiving permission to live out of barracks with a local girl. And his long awaited promotion to corporal arrived.

In January 1820 the Regiment at last embarked for England. Just two soldiers had succeeded in surviving the complete tour. Life had been quite pleasant during the previous year, one of the reasons why half the Regiment, 220 men in all, decided to transfer to other units and thus forfeit their chances of seeing their homes where life, for many of them, had little to offer.

Calladine described the Regiment's leave-taking:

> . . . we had such a number of black women coming alongside, who were left behind, some with three or four children . . . I suppose the 19th Regiment left more children than any regiment leaving the country before, as it was so long in the island, between twenty-four and twenty-five years. Some of them were

grown up, and girls were married, while boys who had been brought up at the
Government School at Colombo were filling respectable places as clerks, or
otherwise had entered the army.[32]

This would have been another cause for so many soldiers deciding to stay.

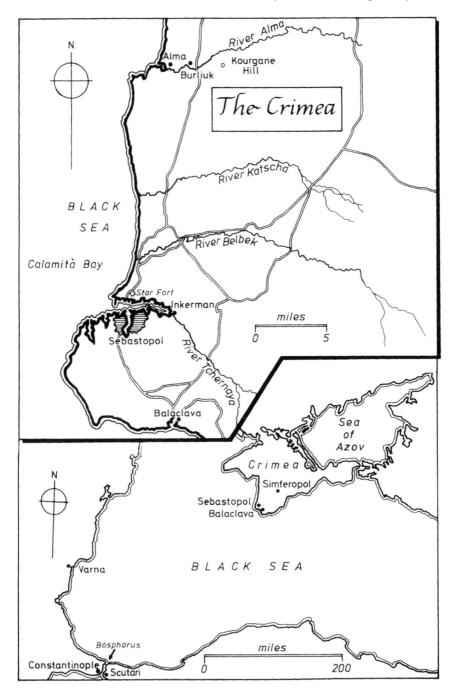

4. From Napoleon
To The Crimea:
1820-56

By early 19th-Century standards, the voyage home was not too unpleasant. Only a single soldier died, and an understanding commanding officer granted Calladine and his friends 'all reasonable indulgences on board, being allowed to play at cards or any other game, and at night we had great amusement in dancing'.[1] Few of the soldiers had thought they would see their native country again, and when the Regiment docked at Gravesend on 18 May 1820, several kissed the ground for joy as they stepped ashore. Some of the youngsters in the band and drums who had been born in Ceylon were so astounded by it all, Calladine remembered, that they went quite wild; the cherry-cheeked barmaids whom they tried to kiss and fondle had to douse them with water to cool their ardour.

It was a mere skeleton of a Regiment of 240 men: half were straight away discharged, some because they were worn out, some seven-year men and some, oddly enough, because they were found to be under height. Within a month, however, the Depot companies joined and a few recruits were enlisted, the command devolving upon Lieutenant-Colonel Alexander Milne. Major Macdonald, promoted Brevet Lieutenant-Colonel in Ceylon, had brought the Regiment back, but Milne had superseded that fine fighting soldier, Macdonald reverting to second in command. The iniquitous purchase system was at work. Milne had purchased his lieutenant-colonelcy in the 19th Regiment on transfer from the 15th Foot, something that the able Macdonald could possibly not afford. However, Milne proved to be a most efficient and popular officer, who quickly knocked his young Regiment into shape.

When the Green Howards celebrated their tercentenary in 1990 (a year late because of operational duties in Northern Ireland), some 2,400 retired and serving members of the Regiment and their wives dined together in one vast marquee after new Colours had been presented. It had happened before. After the Regiment had fired a *feu-de-joi* from a hilltop outside Daventry on 19 July 1821 to celebrate the coronation of King George IV, officers and men feasted together at tables in the barrack square at Weedon. The newly married Pay-Sergeant Calladine remembered the then unique occasion:

The non-commissioned officers and their wives sat at a separate table, the band and drums at another, and the whole of the Artillery and their wives and families at a third. After dinner was over Colonel Milne came to the iron railing that surrounded the elevated space where the officers dined and proposed the first toast, which was 'The King', all the men immediately rising and giving three cheers . . . During the intervals there were toasts and songs sung at the different company's tables, and all was the height of hilarity during the remainder of the day. The Light Company at length went in a body with a chair, and, after mounting the steps where the officers were sitting, secured the colonel and chaired him around the barrack yard, the band and drums playing alternately before him . . . At last the serjeants' wives carried round Captain McDonald's lady.[2]

Gulfs between rich and poor, between officer and soldier, and between the educated and illiterate were then vast. That such a party should have been held says something about the Green Howards, their discipline, and the relationship that existed between officers and soldiers.

***

The Regiment's stay in England was brief (in fact, during the sixty-one years between leaving for Ceylon and returning from the Crimean War, it served there for a mere fifty-four months). At midnight on 24 October 1821 unexpected orders were received to leave for Ireland. After packing throughout the night the companies set out next morning on what was to be an eleven-day march from Weedon to Liverpool, their embarkation port.

Not for the first time nor the last, the Green Howards found Ireland in an unhappy state. During the previous century many Irish Protestants had joined with the Catholics in opposing their country's subservience to England, a movement encouraged by the war in America and which led in 1782 to the Irish Parliament gaining the nominal right to legislate for itself. But this was not enough. The French Revolution was the spark for the United Irishmen movement, its leaders mainly Belfast Presbyterians, but among them that Green Howard officer, Lord Edward Fitzgerald. Aiming to unite Protestants and Catholics into a single independent nation, the United Irishmen sought French aid to such good effect that in 1796 a strong French invasion force arrived in Bantry Bay, only to be dispersed by a storm. The abortive rebellion that followed was ruthlessly smashed by units of the Irish Militia and the regular army, its units often full of Irish Catholics. As always atrocity and counter-atrocity fed one on another.

With this background of recent and savage civil war, it was hardly suprising that the long-established network of peasant secret societies, known by the generic name of 'Whiteboys', controlled much of the countryside, using the customary weapons of murder, blackmail and arson to protest against their appalling poverty and protect themselves against the worst excesses of absentee landlords.

Spread around the country in small detachments, some of no more than four soldiers, the Green Howards guarded gaols, escorted prisoners, searched for arms, carried out four-man patrols and performed such like duties. A hanging shortly after the Regiment arrived required the use of most of the Battalion, as well as police and a squadron of cavalry, to ensure that the three condemned men were not rescued by a vast crowd of spectators who set up an anguished scream as the men were jerked off the cart.[3] Another task was the guarding of revenue officers as they searched for illegal potheen stills, a popular duty for the soldiers who received 1s.8d. extra duty pay for the work, plus bonuses when stills were found.

Lieutenant Henry Rose, later a field marshal and ennobled as Lord Strathnairn, a very Scottish choice of title for a man raised at Duncombe Park near Helmsley, remembered one such foray. A ten-mile night march across peat bogs brought his small patrol to a semi-ruinous cabin, one side of which had been thatched by the sale of the previous year's potheen; the other awaited the proceeds of the current brew. The still was seized amid the shrieks of the wretched family and their friends, but Rose, clearly disliking his work, paid the family well to produce potatoes and milk for his wet and weary men. The next day's work went badly wrong. The searching of tents at a fairground for potheen resulted in the revenue officer having to be rescued from a mob, which then attacked Rose's eight-strong party; knocked unconscious by a stone thrown from a rooftop, Rose was rescued only by his men firing low into the rioters, wounding two of them. The Commander-in-Chief approved his conduct, ruling that 'an officer could not be expected quietly to submit to violent assaults endangering the life of any of his party'.[4]

Calladine tells how on these raids they would find 'the whole family huddled together in a corner on a shakedown or bed of straw. Father, mother and children, and even some of the children grown up, the greater part of them completely naked, with the pigs in another corner and the poultry over their heads'. He was often 'affected by deep compassion for the poor wretches'.[5] Rose and he could hardly have been alone in this, but orders were obeyed, even though a high proportion of the Regiment was Catholic Irish.

The 19th Foot left Ireland in 1826 for a ten-year tour in the Caribbean, where companies moved around between Demerara (now Guyana), Barbados, St. Vincent, Grenada, Trinidad and Tobago. During their time in these lovely but unhealthy islands they lost two commanding officers. The first of these, Alexander Milne, was granted his dying wish that the Colours should be wrapped around his body and buried with him. His successor, the equally able and popular Colonel Hardy, who had served for long years in Ceylon, then succumbed in Trinidad after commanding for seven years. Hardy was, wrote Calladine,

> . . . a great loss to the regiment, for of all the men I ever heard of as Colonel of a regiment no one received so high a character as he did; indeed, he was the father of the regiment, visiting the sick, clothing the children, watching over the men in the barrack rooms to see that they were comfortable, taking care that they changed their linen and clothing when wet, taking women and children when sick to his own quarters, and supplying them with everything that was requisite . . . I have heard several sergeants say that he expended his own money on the regiment in different ways of charity.[6]

The Governor of the Colony wrote about him in similar terms, emphasising that 'his courteous manner and moral example . . . much contributed to establish the character of that corps for all that is correct and gentlemanlike'.[7] For all that, he was as much a disciplinarian as his fellows, content that his regimental court-martial would sentence a soldier to receive 300 lashes. One such, Private Barney O'Ruark, being escorted to his 'breakfast', to use the parlance of the day, managed to throw himself into a vast pond, the product of the barrack cesspools, defying his escort to fetch him out of the waist-deep filth. So amused was Hardy that he promised O'Ruark a pardon if he would emerge of his own volition.[8]

★★★

Calladine himself remained in Ireland with the four depot companies, kept busy on guards and duties as well as training recruits. Life there for the sergeants, especially the married ones, appears to have been agreeable. They had their own mess room, food and liquor was good and plentiful, dances were frequent as were picnics and fishing or shooting trips. Calladine's wife could even keep a servant maid. Against this, the families of corporals and below would lodge in a curtained-off corner of the men's barrack room, a practice that would continue for many years to come. Even a sergeant's wife could find herself on the parish if her husband were not among the fortunate four in a 150-strong draft bound for the West Indies.[9] Death from disease, although not as common as in the West Indies, was an ever-present threat, especially for the children. Mrs. Calladine brought thirteen babies into the world: two survived infancy.

The Depot crossed over to England in 1830, a year that was to be marked by the so-called 'Captain Swing' rural riots, a near insurrection that, as Calladine commented, brought the army some hard patrolling duty; however, he fails to mention whether the Depot was involved.[10] But the following year, when the House of Lords rejected the Reform Bill and the trouble spread to the cities, these Green Howards were continually on the move, criss-crossing the country from one trouble spot to another. At Stamford they quelled a Militia mutiny, and in Yorkshire and Lancashire Calladine found the people generally disaffected, those of Burnley described as 'a vulgar, abusive set of blackguards . . . I never was in a place I disliked more'.[11]

It was back to Ireland for the Regiment in 1837, the depot companies having preceded them. Calladine found many of his old comrades 'reduced, emaciated and sallow-looking', and so much given to drink that a great number were broken to the ranks, their places being filled by promising youngsters from the depot.[12]

Thirty months in Malta followed this Irish tour, before the Regiment moved eastwards in 1843 to the Ionian Islands, were they may have helped introduce cricket to Corfu, a game still played there. It was back to the West Indies again in 1846, but this time only for two years. Canada was the next destination, first Montreal and then Quebec. In Montreal the 19th was part of a large body of troops concentrated there to discourage a movement for annexation by the United States, a crisis that had resulted in a mob burning down the House of Legislative Assembly.[13]

Then, at long last, in June 1851, the Regiment embarked for England.

★★★

Following a year in the West Country, in May 1852 the Green Howards moved to Winchester where they were conveniently placed for the funeral of the Duke of Wellington, one of two regiments ordered to attend. Formed up four deep on the south side of St. Paul's, some of the men found themselves ordered to sling their arms and help push the monumental funeral car up Ludgate Hill, where it had stuck in newly laid gravel after the heavy rain. A spectator remembered groups of officers with green facings plastered up to their knees in yellow slush.[14]

The Regiment had paraded with its Band and Drums. One of the few line regiments to possess string instruments, the band's reputation was then high. Brass bands had not made a general appearance in the army until about 1770, although some had existed a few years earlier. That of the 19th Foot had been formed in Scotland in 1774; the following year it was in good shape with some eleven performers.[15] In the early days bands were dressed according to the caprice of their commanding officers, but by 1830 the white coatees were worn which can be seen in the contemporary painting of the Regiment falling in for the 'Iron Duke's' funeral.

The strong tradition still exists, though unbacked by firm evidence, that Colonel Charles Howard, while on a diplomatic mission to Vienna at the end of 1742, was presented by the Empress Maria Theresa of Austria with three musical scores for his Regiment – a quick, a grand or slow, and a funeral march. No trace exists of the first, but the other two are still in use. It is sad, however, that expert opinion should have determined that the funeral march cannot possibly be earlier than about 1830, and that it may date from the late 19th century; the grand or slow march, on the other hand, is probably contemporary with Colonel Charles Howard's mission.[16]

The drums and fifes of a Corps of Drums were first mentioned in the inspection report of 1758, but the Regiment had marched to their music many years earlier than that. Francis Grose, the antiquarian, wrote in 1786:

> The fife was for a long time laid aside, and was not restored till about the year 1745, when the Duke of Cumberland introduced it into the guards; it was not adopted into the marching regiments, till the year 1747; the first regiment that had it was the 19th, then called the Green Howards, in which I had the honour to serve . . .[17]

The Royal Artillery, as well as the Guards, had brought the fife back into use before the Green Howards, the instrument having fallen into disfavour some fifty years before. There is no record of when the bugle was first used, but it probably coincided with the raising of the light companies.

★★★

It was rare at the time for a regiment to remain in a station for more than a year or so, but the move in 1853 to Weymouth and Gosport came about by what the then Lance-Corporal C.W. Usherwood described as its 'disgraceful disagreement' with the 38th Foot (later the 1st South Staffordshire Regiment).[18] By this time the railways had made these changes of station far less arduous (Calladine had seen the Regiment move by march route seventeen years before),[19] and in July the Green Howards took the steam train for a camp on Chobham Common where they were brigaded for some weeks with two other battalions for training. While there the general made a habit of ordering the Regiment's excellent band to perform when he entertained Queen Victoria and other distinguished visitors.

To concentrate units for manoeuvres in this way, an occurrence depicted in a water-colour owned by the Regiment, was all but unknown at the time. It came about through the Army's unease when Louis Napoleon, the late Emperor's nephew and heir, carried out a *coup d'état* in 1852 to assume Napoleon Bonaparte's mantle as Emperor of the French and exhibit similar aggressive tendencies.

Another and immediate consequence of this renewed threat from France was the Militia Act of 1852, a move to reorganize that now near defunct body. The North Yorkshire Light Infantry Regiment of Miltia, previously the North York Militia and soon to be the North Yorkshire Rifle Regiment, was typical of its fellows. For the past twenty years it had done no training; its 78-year-old adjutant was described as 'inefficient, quite worn out; but intelligent and zealous', while six of the lieutenants had been commissioned prior to Waterloo. Within a year, however, the unit was more than 1,000 strong and soon after the outbreak of the Crimean War it was embodied at Richmond, but not at the Castle, which did not become its headquarters until 1856.[20]

Contrary to expectation, the coming war was to be fought, not against the French but in alliance with them. Arousing vague feelings of fear and mistrust, the vast and expansionist Russian Empire was throughout the century seen as a direct threat to British interests in India; to add to this, recent outrages against Russia's Polish subjects had shocked public opinion. The trigger for the outbreak of this Crimean War was a squabble between Greek Orthodox and Roman Catholic monks in the Holy Places of Jerusalem, then part of the Turkish Empire. A Russian army invaded the Turkish provinces north of the Black Sea, now Romania and Bulgaria, and a Turkish squadron in the Black Sea was sunk. In defence of Turkey, Britain and France then declared war on Russia, both nations united by a fear of Russian penetration into the Mediterranean through the Bosporus and the Dardanelles; at the same time, the French Emperor, well aware how British sea power had furthered his famous uncle's downfall, was determined to avoid Britain's future enmity.

***

On 24 March 1854 two Green Howard companies stationed in the Tower of London, together with the band and drums, formed the escort at the Royal Exchange to the Herald who read out the Royal Proclamation that war had been declared. The Regiment's departure from London, as part of a 26,000-strong expeditionary force, was described by Mrs Margaret Kerwin, one of fifteen wives who sailed with their men:

In the month of April, 1854, about half of the regiment left the Tower of London, under the command of Colonel Unett, and it was a beautiful sight, as all the ladies and gentlemen of London were around the surroundings of the Tower, waving their handkerchiefs and throwing oranges and other things at the soldiers. Old women crying at parting with their sons going to the war, others blessing them, and praying that they may live until they come back, and the regiment shouting and cheering.[21]

After forty years of peace it was not difficult to rouse all classes into a fighting mood.

It was a fine unit of more than 900 long-service soldiers, part of 2 Brigade of Lieutenant-General George Brown's Light Division, that landed at Scutari in May where Lieutenant-Colonel Robert Sanders, who had been travelling abroad, joined them. Like most of the other regiments, it was a well-drilled and disciplined body of men with sound *esprit de corps*. But there was little else. Everything needed to fight an extended campaign was lacking. During the past forty years the Army's scanty administrative services had been run down, there was was no proper reserve, the few staff officers were untrained, the senior commanders were mostly in their sixties and only a handful of the younger ones had known active service.

Nor was there a Marlborough or a Wellington who could weld these individual regiments into a proper army and sort out the chaos. Lord Raglan, the Commander-in-Chief and lately Wellington's Military Secretary, was conscientious, universally popular, hard working, honest and sympathetic. But he had never commanded in the field and for much of the past forty years had filled a desk at the Horse Guards.

In May the Regiment was transported by ship north to Varna where General Brown, a bad-tempered but able martinet and also a Peninsular veteran, proceeded to knock his light division into shape. His brigade and battalion commanders soon gained 'an unenviable notoriety for their harassing tactics'.[22]

But was not long before the combination of heat, insanitary camps, too freely available contaminated fresh fruit and bad red wine began to thin the ranks. Equipped with a water-bottle and haversack, and carrying on her head a small washing bowl for her cooking things, Margaret Kerwin continued to follow her husband, working twelve hours daily washing clothes for the men of No 5 Company and sometimes being bilked of her due. She describes how:

> The men were dying so fast with the cholera, and what they called the black fever, that they had to be buried in their blankets. We moved then up country, and no sooner had we gone than the Turks opened the graves and took the blankets from around the dead men. We were then ordered to bury them without any covering, except the branches and brambles we picked up.[22]

In three months the British and French armies lost 10,000 men, discipline began to break down, and so debilitated were many of the troops that the Guards took two days to march ten miles.[24]

★★★

While Raglan's army lay in Bulgaria 'inactive and literally rotting' as Mrs Kerwin's husband put it,[25] to the north the Turkish armies had repulsed the invading Russians by their own efforts, thereby averting the danger to Constantinople. Work, therefore, had to be found for the Allied armies, despatched so far at such great expense. The result was that the British and French governments directed them to attack Sevastopol, the great Russian naval base at the tip of the Crimean peninsula, the capture of which would make easy the destruction of the Russian Black Sea Fleet. So with Sevastopol as their objective, the Allies embarked once again at the beginning of September, leaving behind their many sick, among whom was Margaret Kerwin, whose husband ran out from the ranks to kiss his apparently dying wife farewell.

On 14 September the Green Howards were landed on the open beach at Calamita Bay, twenty-seven miles to the north of their objective. All night it rained and the tents were not yet ashore. Usherwood, now a sergeant, with two others built themselves a shelter of dried grass, covered by a blanket; others, officers and men alike, who had slept in the trampled mud, paraded the next morning in an even filthier state.[26] Knapsacks had been left aboard ship (later to be plundered) and only a blanket and greatcoat were carried rolled around a few necessaries. The only transport the Regiment could boast was a pony commandeered from a Tartar village and allotted to the surgeon to carry his two small medical panniers.

The rest of the Army was just as short of transport. Everything contrasted unfavourably with the French whose organization had been tempered in recent North African colonial wars and whose equipment was the envy of their British allies. To French officers the British appeared as enthusiastic amateurs, gallant but inexperienced. Not that French general-ship was much superior to that of their allies. Nor, of course, did any type of joint command exist: battle planning was a matter for discussion, sometimes but not always amicable.

Four days were needed to land the rest of the Allied armies and their stores, and it was not until 19 September that they began their southwards march towards Sevastopol, twenty-seven miles away. On the right or seaward flank were some 7,000 Turks, next to them moved the French and on the left the British, each some 27,000 strong. Parallel sailed and steamed the fleets. They were armies operating virtually in the dark. Nothing was known about the strength of the Russians and little more about the topography of the Crimea.

William Howard Russell, the first civilian war correspondent, described the mass of gaily uniformed men, covering several square miles of countryside and descending the ridges, rank after rank, with the sun playing on the forests of glittering steel.[27] Closer at hand, the picture was not so glamorous. Sweltering in full dress uniforms, unfit men soon finished the water in their wooden canteens, and many dropped by the wayside from cholera and heat stroke. The way was marked also by discarded leather stocks, a vicious article of dress still worn by most regiments but packed away by the Green Howards some months ago in Varna.[28] Ahead the smoke from burning farms and hamlets showed that the enemy were alert to the Allied advance. Soon bodies of Cossack cavalry began to harass the leading troops.

That night, as the men munched their salt beef and biscuits, they could discern the enemy positions far away across the River Alma, clearly defined by twinkling camp fires.

What happened to the 19th of Foot on the following day is not easily unravelled although there were eye-witnesses in plenty, among them the second in command, Brevet Lieutenant-Colonel Unett, Captain Lidwill, the 18-year-old Ensign and Adjutant Cardew, Private John Kerwin and Sergeant Usherwood, all of whose accounts of the battle not surprisingly differ. The Duke of Wellington when writing to a hopeful chronicler of Waterloo, likened the history of that battle to a history of a ball, 'No individual can recollect the exact order in which, or the exact moment at which, they [particular incidents] occurred, which makes all the difference as to their value or importance'.[29]

The Alma, a winding river, was fordable in most places that September. Rolling hills, similar to the wolds of what was the East Riding of Yorkshire, overlooked it from the south; along their smooth rounded slopes and on top of their flat plateaux were arrayed the solid masses of the Russian infantry and cavalry, some 40,000 men, ready to crush the invaders.

That morning, the British, after making a scanty breakfast on their few scraps of food, made a slow start. Everyone was weary, orders were confused and the Allied command (described by one Green Howard present as a 'two-headed monster'[30]) was determined to avoid friction and thus found it near impossible to agree upon a joint plan of attack. Not until ten o'clock did the Green Howards move off on the right flank of 2 Brigade which was drawn up on the left of the Light Division which lay in turn on the left flank of the Allied infantry. George Lidwell mentioned:

A curious stillness and silence pervaded the army. No bugles were allowed to sound, and unless the neighing of a horse or the rocking of the wheels of a gun on the soft ground, no sound escaped our line . . . The ground traversed was of a deep clay soil covered with a coarse dried-up vegetation with an odour like thyme.[31]

At about eleven-thirty a halt was called while the Allied commanders held a further conference on top of a knoll in full view of the British troops. On the hills ahead of them, scarred by the breastworks dug to protect the enemy batteries, could now be seen the solid grey Russian columns. The most prominent of these batteries, sited on the flank of Kourgane Hill and mounting a dozen guns, was to be known to the British Army as The Great Redoubt. Further batteries covered each flank. For these unblooded troops it was a daunting sight.

At one o'clock the armies began once again to move forward, the French to assault the hills near the mouth of the river. Soon the Green Howards began to come under fire from the batteries above them, one round shot smashing a grenadier's leg and another slicing off an officer's calf (gangrene

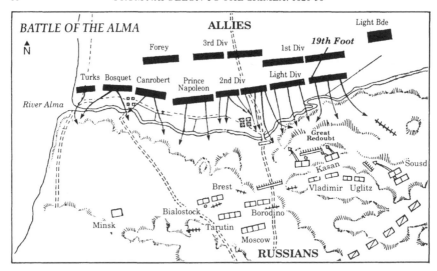

later killed him). With superb precision the Regiment deployed into line, but much confusion ensued on the right where other units had converged. Waiting, however, for the French assault to succeed before assaulting the Russian position, the British Army halted, and the men, lying down in the ranks, suffered for ninety minutes a hammering from round shot, grape and the musketry of the Russian skirmishers.

At last came the order to advance. Scrambling towards the river through vineyards and over banks and walls, and losing men all the time, the Regiment at last reached the water's edge, but in some confusion, most of the battalions on their right having become irretrievably mixed. Into the river everyone plunged, weapons held over their heads and ammunition pouches kept dry under armpits, some men only ankle deep but others sinking into deep holes in which some drowned. Kerwin, a cautious non-swimmer, first hesitated on the brink before plunging into the water two steps behind his officer, ready to retreat if the subaltern should disappear beneath the surface. Clambering out he saw the ensign carrying the Regimental Colour shot dead through the forehead and a round shot cut the bowels out of four of his comrades.[32]

Having crossed the river the brigade commander, lacking clear orders, decided to halt in order to protect the Army's left flank from a possible Russian cavalry attack. The Green Howards, however, had by then become so intermingled with General Codrington's brigade on their right that they failed to receive the order and swept up the smooth, 500-yard-wide glacis in the face of the fire of the guns of the Great Redoubt. The Regiment was still in some sort of order, but Codrington's units now consisted of disorganized groups of soldiers, urged forward by whatever officers were at hand, among them Codrington himself, in action for the first time and bellowing from horseback at those around him. Nearly level with the Great Redoubt, the Regiment saw ahead a two-battalion Russian column dressed in what Lidwell described as 'their strange drab coats and ugly heavy-looking helmets'.[33] A newspaperman, Mr A.W. Kinglake, remembered how the 19th together with some companies of the 23rd:

> In their English way, half sportive, half surly, our young soldiers seemed to measure their task; and then – many of them still holding betwixt their teeth the clusters of grapes which they had gathered in the vineyards below – they began shooting easy shots into the big, solid mass of the infantry which was solemnly marching against them.[34]

To the Russian officers on their vantage points above the valley, the British, some in wavering lines but most in rough groups, appeared to be to be nothing more than advanced skirmishers. It seemed presumptuous that they should be attempting to assault those solid columns. But those columns, reminding Lidwill of a Derby Day crowd, began to waste away as single bullets from the 19th Regiment's newly issued Minié rifles penetrated two or three Russian ranks. Soon the grey mass swayed ponderously away to the east and disappeared into dead ground.

Meanwhile men from Codrington's brigade, with them a large group of Green Howards, probably No 6 and the grenadier companies which were on the right flank, had assaulted the Great Redoubt through storms of grape and canister, taking heavy casualties but driving out the Russians.

Codrington, mounted on a grey Arab and leading from the front, had in some miraculous manner survived the assault, unlike the mounted Green Howard officers, all of whom had been hit. Of the five regiments who assaulted Kourgane Hill, some 2,000 men reached the summit; as happens at such times, they then clustered in and around the captured Redoubt, regaining their breath, tending the wounded and congratulating themselves upon still being alive. Codrington, however, quickly took control, placing men and units in place ready to repel the inevitable counter-attack from the 10,000 Russian infantry grouped around the position, most still unscathed.

The Russian counter-thrust was not long in coming. Soon one, and then a second mass, each some 3,000 strong, began to move slowly and steadily down the hill towards the Redoubt, now under fire from another Russian battery. There followed a typical muddle. First, an unknown mounted officer ordered a bugler of the 19th Regiment to sound the 'Cease Fire', crying out that the approaching troops were French. Then the same or some other officer, not a Green Howard, ordered the 'Retire' to be blown. The result was inevitable. At first many queried the order, but soon a general retreat started, for the most part unhurried, men carrying their wounded with them and turning from time to time to fire at the enemy once again in possession of the Redoubt.

Behind the Light Division, and instructed to support it, waited the 1st Division, one of its brigades composed of Highlanders and the other of Guards. When at last ordered forward, its commander, the Duke of Cambridge, did not rush. In fine style the Guards moved down to and across the river under a murderous hail of fire, halting to dress their ranks before starting to climb the slope littered with nearly a thousand Light Division dead and wounded, but too late to relieve its now retreating units, the survivors of which passed through their ranks. One confused battalion broke under the heavy fire, following the light division back to the river. But threatened by the cool precision of the rest of the Guards and the Highlanders and hammered by the British guns, the Russian columns abandoned the battle and began to retreat towards Sevastopol, unhindered by the British cavalry which failed to pursue the now beaten enemy.

After Colonel Sanders had been badly wounded in the leg, Unett – also hit but not so seriously and with his face smeared with his horse's blood – rallied the Green Howards and marched them back up to the top of the hill behind the Guards. There they bivouaced for the night after looting discarded Russian knapsacks. In this way Sergeant Usherwood collected four loaves of black bread, tasting of chopped straw; he then fell asleep in his soaked clothes with an unchewed chunk in his mouth. More valuable booty were the seven fine drums of the Vladimir, Minsk and Borodino Regiments, still treasured by the Green Howards.

It took two days to clear the battlefield of the last of the wounded, bury the mounds of dead and collect the discarded Russian equipment. Losses to the Green Howards had been heavy, nearly 250 officers and men, an eighth of the total British casualties, greater than in any other unit except the 33rd. But as Usherwood commented, the figure would have been still higher if all those who had been contused or slightly wounded, himself among them, had reported themselves.[35] Names such as Doyle, Quinn, Doherty and Kelly are scattered throughout the casualty list. Years before, Calladine had commented that half the Regiment was Irish Catholic, a ratio that seems to have held steady for many more years.

Each year the Green Howards still celebrate 20 September, the anniversary of the Alma battle, as their Regimental Day, trooping the captured Russian drums when opportunity permits.

***

The Allies started in pursuit of the Russians on 23 September in what can best be described as a ponderous manner. Marching through attractive countryside, the ground littered with equipment and stores abandoned by the enemy in their headlong retreat into Sevastopol, the men's morale was high despite continuing losses from cholera. There was fruit to be picked and pleasant villas to be plundered; the 19th Foot was specially fortunate in rounding up a complete Russian convoy, its wagons full of every sort of luxury.

Circling around to the east of Sevastopol, the Allies occupied the harbour of Balaclava, henceforth to be used as the main base. They then invested the eastern suburb of Sevastopol, the key to the Russian defence, as its loss would render the adjacant naval base untenable.

At first the weather was pleasant and by 9 October tents had arrived. But, only too quickly, everything began to go wrong. For the Green Howards their sufferings were to be similar to those of their successors in the First World War, a year of arduous trench duty and working parties, the grinding toil interspersed with bloody attempts to breach the Russian defences.

The Regiment was not involved in the first major Russian sortie which culminated in the disastrous charge of the Light Cavalry Brigade outside Balaclava, nor did they take a major part in the famous battle of Inkerman on 5 November. This second assault by the enemy, now outnumbering their opponents by two to one, all but succeeded. In this 'Soldiers' Battle', as it was to be known, small bodies of British troops, often cut off one from another, usually fighting hand to hand in the thick fog that enveloped the battlefield, held off overwhelming enemy numbers to such effect that in the end the Russians pulled back. With most of the 19th on duty in the trenches, only three companies could be collected to help reinforce the scanty British force struggling on the Inkerman plateau, and although they were continually under fire, they were not committed, losing no more than five dead, including the Regiment's sergeant-major.

By the time of Inkerman that first terrible Crimean winter had set in. Clothing had worn out; men, sometimes spending three nights out of four on trench duty and frequently under heavy fire, were often soaked to the skin in the sub-zero Russian weather. On 14 November a hurricane devastated the camps, sweeping away the tents and much else, leaving the sick and wounded lying unprotected in the snow and mud. Although there was food in plenty at Balaclava, there was little transport to carry it the seven miles to the troops. Of fuel there was practically none, and so ineffective was the commissariat that coffee beans were issued green. The sole comfort was rum. In such conditions of cold, wet, filth, overwork and semi-starvation, disease again spread, scurvy compounding the cholera and fever.

Only pitiful, half-trained boys of sixteen and younger arrived to replace the fine long-service survivors of the Alma, and these often perished as soon as they landed. Of the draft of 102 men which joined the 19th on 21 November, the greater part, Usherwood noted, 'died from disease 'ere the end of the month had expired',[36] an exaggeration possibly, but an indication of the horrible conditions. Some regiments ceased to exist. Others, better commanded, just survived. Among the latter were the Green Howards, its colonel now Thomas Unett, a man of great character who could speak his mind. Nevertheless, the ranks were so thinned that it became near impossible to find the numbers needed for the trench duties. The 'ragged, dirty, and bootless' men, 'melting away from want of food', had become weary of life, as a Green Howard officer later admitted. Even Unett was heard to complain aloud that Sevastopol would never be taken because there was not a general up to the job.[37]

By early spring, a little had been done to improve conditions, Usherwood noting after two fine March days, that:

> The men are now beginning to pick up a little, a smile or two have seated themselves upon their countenance displacing the forlorn abjecture that so long had possessed himself of this seat of horror, better food, more clothes, less disease and finer weather rousing the almost worn out soldier into vigour-hood.[38]

The resilience of those Green Howards was remarkable.

At such a juncture it was typically inept of the divisional commander, Sir George Brown, to issue orders that men must wear stocks and that any lost shakos would be replaced at the wearer's expense.[39]

As the siege warfare continued into the following summer, the gallantry of two soldiers, Samuel Evans and John Lyons, attracted special attention. Two years later, when the Victoria Cross was instituted, both men were awarded the decoration. So coveted in the future, the award at first lacked prestige. Many officers disapproved because they held that the estimation of a man's comrades was an adequate reward for deeds of valour.

The core of the Sevastopol defences consisted of two strongpoints – the Malakoff, opposite the French, and the Redan, half a mile to its south-east, faced by the British. On 18 June 1855, after nine months of preparation the Allies had accumulated the resources to mount the first deliberate attack, but this the Russians threw back. Casualties to both sides were heavy. As part of the British reserve the Green Howards were not involved.

For twelve more weeks the Allies pushed forward their trenches and intensified their bombardment, suffering all the time a steady drain of casualties, until they were ready to try again. In the culminating and near crushing final three-day cannonade, 800 pieces of ordnance poured 100,000 rounds into the Russian defence works. Then, on 8 September, the French assault troops bounded across the mere twenty-five yards between their forward trenches and the Malakoff, thrust the Russians out, and then held it against a series of counter-attacks.

The British had much farther to cover, some three hundred yards of open ground, sapping forward beforehand through the near solid rock having proved slow and difficult work for British troops in any case reluctant to use their picks and shovels.

The plan of attack was for a 1,000-strong storming party to advance behind covering groups of marksmen and others carrying bags of wool to fill the twenty-foot wide ditch, and ladders to scale the 30-foot-high wall. Behind this storming party were 1,500 strong support groups, with reserves of double that size waiting on call still farther back.

The 19th Regiment, now mustering only 438 all ranks, formed part of the supports. Pressing forward on the heels of the stormers, a few of whom despite heavy losses had managed to clamber up the ramparts and enter the works, the Green Howards crossed the ditch and took cover behind the battered traverses. So far they had done well, advancing in fine order through murderous fire. But now the effect of months of static fighting was to be felt: the young recruits, lacking a proper stiffening of old soldiers, could not be urged on any farther. Men crowded into the forward defences, only to be slaughtered by the intense enemy flanking fire. Most of the senior officers had been hit, but their juniors, near children themselves, did their gallant best to control the intermingled units. The bravery of one of these boys won national acclaim. The 17-year-old Lieutenant Dunham Massy, commanding the Grenadier Company, first into the Redan and badly wounded, stood up on the wall to encourage his men forward. For the rest of his life he was to be known as 'Redan' Massy'.

Above: 'Redan Massy', the hero of the storming of the Redan. The newly created Victoria Cross did not come his way, but at the age of 18 he gained accelerated promotion to captain, a far more satisfying reward.

In the end, the reinforced Russians massed in overwhelming numbers to thrust out the British with their bayonets, savage combat in which the swords of the officers were shown to be hopelessly ineffective weapons. Driven back into the ditch, already choked with dead, dying and leaderless men, the British were a helpless target for the Russians above them. Musket balls, grenades and even rocks rained down. The result was inevitable. The survivors could take no more and ran. Fortescue described the battle as 'disaster and disgrace'. Disaster it certainly was, but for the Green Howards it was no disgrace. In an ill-planned operation, untrained youths had crossed that fire-swept slope and held their ground for an hour before they broke.

Lieutenant-Colonel Unett, acting as brigade commander, had led forward the leading troops of the light division. Beforehand he had tossed a coin with Colonel Windham of the 2nd Division to decide who should lead the assault. Winning, he vowed that he would be the first man into the Redan. He was not. Instead he was one of the earliest to fall, mortally wounded. Margaret Kerwin, who had recovered and rejoined her husband, wrote:

> I heard the Colonel was wounded, and I went down to meet him, the shot and shell flying in every direction. I saw the Colonel and took out my handkerchief to wipe his face, and he asked after my husband. He did not live many days after that, and both myself and my husband went to the funeral. It was a sad day for both of us to lose Colonel 'Daddy' Unett, as he was called in the Regiment. We buried him on the hill outside Sebastopol, where all the soldiers were buried. It was a very small funeral, as most of the men were on duty.[40]

It was not surprising that so few men could attend their Colonel's funeral. Nearly half the Green Howards who had started across that open ground were dead or wounded.

Although the Russians still held the Redan at the end of the day, the loss of the Malakoff had made Sevastopol untenable. Quietly and skilfully the Russians evacuated the city. The attrition of the siege had cost them 30,000 men, and they lacked taste for any further fighting. Afterwards the war drifted slowly and uneventfully towards its end, but nine months were to pass before the Green Howards took ship for England. No record has come to hand, but few of those who had landed in Turkey two years before could have returned home with their Regiment. The killed and wounded numbered nearly 700. Nearly 317 others were dead of disease.[41]

5. QUEEN VICTORIA'S SMALL WARS: 1856-1902

The Regiment's stay in England was brief. Just a year after landing at Portsmouth, it was again taking ship in July 1857, this time for India, its task to help quell the Great Mutiny. In this and similar small wars aimed at extending, consolidating or defending Queen Victoria's realms, the Green Howards were to be kept busy until the British Empire reached its zenith at the turn of the century.

Succeeding her uncle, William IV, in 1837, Queen Victoria had married three years later and was to be widowed in 1861. In Rudyard Kipling's famous lines:

Walk wide o' the Widow at Windsor,
For 'alf o' Creation she owns:
We 'ave bought 'er the same with the sword an' the flame,
'An we've salted it down with our bones.[1]

'The sword and the flame' brought misery to some, but the benefits of peace, security and civilization to many more, as empires sometimes do. And, as always, it was disease rather than the sword and the bullet that salted the bones of so many young Green Howards in so many corners of the globe.

During the months when the Green Howards were making their slow passage to India, the Mutiny had been broken, leaving the Regiment little to do but guard prisoners in Bengal and then pursue fleeing sepoy bands into the jungles of the Himalayan foothills as far as the borders of Nepal, to which the survivors had fled. It was hard marching, with all the privations of active service but little of its excitement. Soon the last remnant of the Mutiny was crushed, the administrative powers of the East India Company were transferred to the India Office, and the Queen assumed the title of Empress of India.

The Regiment's subsequent garrison life on the plains of India was described in a lively fashion by Edward Spencer Mott, who joined in 1862 as an 18-year-old ensign at Mian Mir near Lahore in the Punjab. A cheerfully irresponsible lad, ill health forced him after seven years to forego a military career for first, the stage and then sporting journalism in which calling he achieved fame under the pen-name of 'Nathaniel Gubbins'. His amusing book, *A Mingled Yarn*, describes how high-spirited young officers tried to alleviate the heat, boredom and ever-present threat of disease that afflicted life in the Indian plains.[2] His description of the fear engendered by the cholera epidemic in the summer of 1862 is compelling reading; fewer than

Above: Officers of the 1st Battalion at Peshawar, 1866.

Left: General Sir Robert Onesiphorous Bright, GCB, Colonel of the Regiment 1886-96.

Top right: The XIXth Foot leaving the Tower of London on 23 April 1854 to embark for the Crimea. Watercolour by R. Ebsworth. Margaret Kerwin described the scene.

Right: The Battle of the Alma. Oil Painting by Peter Archer.

Left: Colonel Thomas Unett. Oil painting by Reuben Sayers. In the Crimea he was known to his men as 'Daddy' Unett because of his concern for them.

Below: The 19th Regiment storming the Redan. Scaling ladders and fascines for the ditch can be seen. Watercolour by Orlando Norie.

50 per cent of the 144 men, women and children of the Regiment who caught the disease recovered.[3] He wrote also of his being nearly sick at his first 'flogging parade', when he watched a soldier receive fifty lashes;[4] it is a reminder that this brutal punishment was not finally abolished until 1881, although its use in the sixties was becoming increasingly infrequent. Attendance at public executions Mott found equally unpleasant.

Much of the book, however, is devoted to the light-hearted way of life of a collection of young officers, described by an Irish doctor as 'The jolliest lot o' divils in the world, me bhoy, but not a pice among them. Sure, all their gold is on their mess-waistcoats . . .'[5] (perhaps little has changed: the waistcoat still carries more gold than that of any other line regiment). But dress at the time was by no means uniform, as can be seen from a number of old photographs, nor was it especially smart, creased and baggy clothes seemingly the norm. Upon the abolition of the flank companies – the grenadier and the light – soon after the 19th arrived in India, two photographs were taken of bearded members of the grenadier company who had fought in the Crimea. All wear their medals, some on the right breast, some on the left.[6] Although polo, mentioned by Mott a little contemptuously as 'horse hockey', was only just coming into fashion, the life of these young men was centred largely on the horse, with racing and hunting the main outlets. Mott hunted at Jullundur with the first recorded Green Howard pack of imported English fox-hounds, its quarry the jackal and its master Colonel Robert Onesiphorous Bright (later a general and Colonel of the Regiment). By an odd coincidence, it was at Jullundur in 1942, in the middle of the Second World War, that the Green Howards Hunt was finally dissolved when the 2nd Battalion was moved to the Frontier station of Razmak.[7]

In 1868 this peaceful but boring existence was interrupted when the 19th were called upon to join an expedition against the Hazara of the Black Mountains, the Regiment's first experience of warfare on the North-West Frontier, that mountainous region inhabited by fiercely independent tribesmen, almost impossible to administer and partial to raiding down into the soft plains. An attack on a much resented police post was the cause of this 10,000-strong punitive expedition being assembled.[8] At the time little was known about either the Black Mountains or their people.

The Green Howards were physically fit, made so by a summer of road-building in the foothills. They needed to be. The battalion concentrated at Abbotobad on 13 August, one wing covering sixty-five miles in fifty-nine hours in the middle of the hot weather without a man falling out (by another coincidence, the 2nd Battalion marched up that same Rawalpindi road in similar sweltering heat just seventy-four summers later). Another, but far less fit, British battalion lost thirty-eight men from heat stroke on the road, eight of whom died. An Indian unit of Sappers and Miners marched 600 miles in twenty-nine days to get there. One participant recalled that there was an immense amount of hardship and discomfort in the course of the campaign, but little fighting.

While waiting for the balance of the force to arrive, the Regiment had a month to practise the recondite but complicated skills of frontier warfare, the bedrock of which was the picqueting of the heights on either side of the line of advance or withdrawal, covered by the guns of the mountain batteries, in order to allow the long column of soldiers and their mass of camp-followers to advance free from the dangers of ambush. It was all very tortuous.

The objective, the summit of the Black Mountain, involved a climb of 10,000 feet through rugged country, much of it dense forest and tangled shrub-covered ravines. It was here that Green Howards first learned how these Muslim tribesmen, dressed in their dirty rags, could make themselves invisible among their native rocks, and the speed of reaction needed to deal with the ambush party which would break cover without warning, stabbing and cutting with dagger and sword.

The Green Howards formed part of the leading brigade, its commander their own colonel, now Brigadier-General Bright. Bivouacking on a ridge about half-way up the mountain, the force protected itself behind stone sangars, entrenching tools being then unknown and, in any case, of little value in such rocky country. Soon the tribesmen closed in and, from their perfectly concealed fire positions, began to make life unpleasant. Fortunately the accuracy of their ancient muskets was poor, only a small number of Indian and Gurkha troops being hit (when modern rifles were acquired, frontier warfare became much more hazardous). From time to time the enemy broke cover and made gallant rushes towards the encampment, but every time they did so the controlled fire power of the defenders stopped them.

Two days after setting out the column reached its objective. The enemy, who had assembled in large numbers on the upper slopes of the mountain, their drums beating out their threatening rhythm and their flags flying, dispersed at the sight of the Green Howards and 2nd Gurkhas advancing remorselessly towards them. It was well that they did so because the expedition's commander 'had never, in the border hills, seen such a naturally strong and defensible position as this peak'.[9] By then, clans were already tendering their submission, and soon all had done so, although some tribesmen continued to harass the column during its subsequent withdrawal.

The terms imposed were not exacting, the tribes having been taught that they were not impregnable. The result was that no further serious trouble occured in these Black Mountains for the next twenty years. Total British and Indian casualties numbered five killed and twenty-nine wounded, none of them Green Howards. It had been a typical little Frontier expedition, a fillip to the training of all ranks and a break from the tedium of garrison life.

***

Meanwhile a second regular battalion had once again been raised for the Regiment, its third incarnation. It was exactly one hundred years since its last predecessor had ended its brief two-year life before taking on the guise of the 66th Foot. At Exeter on 9 March 1858, 350 men from the depot companies of the 1st Battalion were to form the nucleus of this new unit, its establishment fixed at forty-five officers and 1,081 other ranks, its weapon the new Enfield pattern muzzle-loading rifle, sighted up to 1,000 yards and fairly accurate at half that distance. The commanding officer and most of the subalterns were Green Howards, but most of the senior officers were from other regiments; the latter were in general a battle-hardened lot, veterans not only of the Crimea and the Mutiny, but small wars fought in places as diverse as New Zealand, South Africa, Burma and the Punjab. When the recruiting parties, despatched in every direction, had gathered their harvest, the battalion was found to have a ratio of ten Englishmen to six Irishmen and one Scot. Only one in five of the soldiers was a Roman Catholic:[10] famine and emigration had so reduced Ireland's population that it was no longer a bottomless reservoir of recruits.

This raising of a fresh 2nd Battalion had been part of a twenty-eight battalion expansion of the infantry. The reasons for this increase were diverse. The Crimea had exposed the Army's weakness, and the Mutiny the need of a far larger number of British troops in India, their numbers being increased from one-ninth to one-quarter of the total garrison. Trouble threatened from a number of places in the rapidly expanding British Empire – even Canada was menaced from across her southern border by a private army of Fenian expatriate Irishmen. Over all loomed the burgeoning power of Prussia, aggressor against Denmark in 1864 and Austria two years later, preliminaries to her victory over France in 1871 and the unification of Germany under the Prussian monarchy. Britain's seapower could still guard the country against invasion from continental Europe, but there was a growing fear that an army might be needed across the Channel if she were to have a say in Europe's future. And now the threat was from this new Germany rather than from her old enemy, France, even though as recently as 1859 Napoleon III had reawakened fears of French aggression by invading northern Italy.

In the Army of the mid-19th Century life was hardly stable. During the first five years of its existence the 2nd Battalion was subject to a succession of rapid changes of station in England and Ireland, nine in all, before it embarked in August 1863 for Burma, then under the Government of India. A farewell report in the *Irish Times*, revealing of the social attitudes of the era, suggests that this must have been a rather unusual unit:

> In the first place, for so young a battalion, their smartness and soldierlike bearing is remarkable; in the second, the wonderful unanimity prevailing between the officers and men shows how much better it is for gentlemen and officers to recollect that they are men, and that the privates and non-commissioned officers are human beings and not a machine. In the case of this regiment such has been the principle, and the consequences show how much the soldier values the kindness of his superiors; and thirdly, without exception during our knowledge of military men we have never come across so well-behaved a set of non-commissioned officers and men; in fact, in the ranks of the 2nd Battalion, 19th Regiment, there are many men fit by their education and manners to fill any station in life.[11]

From Burma the Battalion was to return to India proper five years later and there it remained until 1873. As the 1st Battalion did not sail for England until 1871, both battalions of the Regiment were together in India for three years, from every standpoint an unfortunate arrangement.

Up-country Burma where the 2nd Battalion was first stationed, was far from salubrious. Mott was posted there in 1867 after purchasing his promotion from ensign to lieutenant, a step that involved a move between the two battalions. For him it was 'a barbarous country, at the back of beyond, peopled principally with dacoits and Chinamen, chronic rain during nine months of the year, a land reeking with snakes, miasma, mammoth centipedes, and creeping things . . .'[12] But, then, he had arrived under a cloud, facing a general court-martial after seven months' absence without leave.

In 1871 this archaic system of promotion by purchase was at last abolished. With officers buying each of their steps in rank on a rising scale – the cost to Mott would have been £250 – it is difficult to understand how the British Army had, over the years, managed to function as well as it did; the system's apparent sole merit was that an able young man, given the necessary resources, could achieve high rank at an early age, Wellington being the prime example. The ending of purchase brought a multitude of benefits, but did nothing to speed up promotion: in the seventies, seventeen Green Howards subalterns, some 'unmistakably grey' averaged more than fifteen years' service each; one had to wait more than twenty years for his captaincy.[13]

Much had been done both during and immediately after the Crimean War to correct some of the worst of the abuses that afflicted the British Army, this abolition of purchase being merely one of the measures taken by Edward Cardwell, the great reforming Liberal Secretary of State, to create a coherent military system. A year earlier, the twenty-two year enlistment (for many a life sentence) was changed to six years with the colours and six with the reserve, thus removing a major deterrent to recruiting and creating a viable regular reserve. Pay was also increased, a free ration of meat and bread was issued for the first time, and flogging was abolished except on active service.

Allied to these reforms was Cardwell's introduction of the 'linked-battalion' system whereby regiments were to consist of two regular battalions, one serving at home and one overseas, a simple and reasonably trouble-free reorganization for the senior twenty-five line regiments which already possessed two battalions but for others a distressing business that caused long-lasting bitterness. To improve recruiting militia and regular battalions were linked in territorial districts, and regimental depots were set up in the middle of these districts. For the first time regiments had permanent homes.

For the Green Howards the outcome was the creation of a four-battalion regiment, the two militia units, previously the 5th West Yorkshire Militia and the North York Rifles, being numbered as its 3rd and 4th Battalions. Eight years earlier the Depot had been established at Richmond, that beautiful and historic little town, dominated by its Norman castle over-looking the River Swale, and for long the centre for the North York Rifles.

Allied to these changes was the abolition of regimental numbering, the Regiment becoming The Princess of Wales's Own (Yorkshire Regiment). And, as with other English line regiments, except those termed 'Royal', facings were changed to white. Needless to say, the loss of the green facings, not to be restored until 1899, was as highly unpopular as the abolition of the '19th'. Five other regiments had 'Yorkshire' in their title, and there was to be much confusion until at last, in 1920, the historic nickname of 'Green Howards' became the official designation of the Regiment. In 1881 also, the cipher of HRH Alexandra, Princess of Wales, interlaced with the Danish Danebrog and surmounted by a princess's coronet was introduced as the badge for the buttons and appointments of officers. At this time also, 'The Bonnie English Rose', in use since 1868 as the Regiment's quick march, was given official approval.

The Regiment had first met this beautiful Danish princess, the wife of the future King Edward VII, at Sheffield on 17 August 1875. There she presented new Colours to 1st Battalion and consented, when the ceremony ended, to the Regiment being designated 'The Princess of Wales's Own'. It was the start of the Regiment's close connection with the Royal Houses of the Kingdoms of Denmark and Norway. In 1914 the by then widowed Queen Alexandra became the Regiment's first Colonel-in-Chief, to be succeeded, in 1942, by King Haakon VII of Norway and later, in 1958, by his son, King Olav V, Queen Alexandra's grandson and great-grandson to Queen Victoria. Then, in 1992, King Harold V was to succeed his father King Olav.

Below: Presentation of Colours by HRH Alexandra, Princess of Wales. Her husband, the future King Edward VII, is on her right. Published in The Graphic, *1875.*

The Colours buried at Demerara with Lieutenant-Colonel Milne are the first about which anything definite is known, although it is recorded that new Colours were either issued or presented to the 1st Battalion in 1759, 1773 and 1780. Those carried at the Alma, the last to be taken into action, had been received by the 1st Battalion at Cork on 21 March 1837, but there is no record of any ceremonial presentation. In fact until mid-Victorian days, Colours would seem to have been treated with far less punctilio than later became customary. When the 2nd Battalion received its new Colours two years after it had been re-raised, they were presented by a mere major-general, the commander of Aldershot District.[14]

A footnote to life at this period was the 1st Battalion's introduction to football, first played at Gosport in 1872, a rough and ready sport, in which half-a-dozen or so officers joined, one of whom had his Sandhurst colours. It was an odd game, played with a round ball, in which hacking and tripping were considered normal, and in which the sole rule was that the ball could be caught only on the first bound. Nevertheless, civilian opponents could be found.[15]

The return of the 2nd Battalion from India in the spring of 1877 was the signal for the 1st to leave for Bermuda that autumn. Three years later history repeated itself when, as had happened twice before, the tropical heat of the West Indies was once again exchanged for the bitter cold of a Canadian winter. All ranks had anticipated the move with 'great jubilation'. For the soldier the grass on the other side of the fence is always that bit greener; the language when they landed at Halifax, Nova Scotia, in white helmets and summer clothing in 12 degrees of frost has not been recorded. Eventually greatcoats, mufflers, otter fur caps and long boots became the order of the day, with fur-trimmed blue frock coats and gauntlets for the officers, specimens of which survive in the Regimental Museum. Partly because of the bitter cold, desertion to the United States became commonplace, one of the unit's tasks being to find the guard to search outgoing trains. It was said that one half of the Battalion was employed to catch the other. The Band's first cornet player, who left wearing his false front teeth was extradited only because it was possible to charge him with stealing two government teeth. When a newly arrived commander prohibited shaving as a counter-measure to the cold, those too young to produce respectable beards were sadly embarrassed. Life in the summer was far more agreeable; fishing was then the main sport, lobsters being speared on the surface by wooden forks which the fish clutched when touched.[16]

Early in 1884 the Battalion sailed for Malta, a staging-post on the way to active service in Egypt. Two years earlier a large British force had invaded and taken control of that country. The security of the Suez Canal, constructed in 1869, had been threatened when an Egyptian officer, Colonel Arabi Pasha, led a nationalist revolt against the foreign financial domination of his country, following which a number of Europeans were massacred in Alexandria.

Inevitably the occupation led to the British taking a hand in the affairs of the Sudan, that vast and almost trackless desert country. When the Mahdi (or Messiah) raised the Sudanese against their oppressive Egyptian rulers, William Gladstone, the Liberal Prime Minister, sent General Gordon (who had previously served in the Egyptian administration of the Sudan) to evacuate the Egyptian garrisons and officials. The outcome was that Gordon was besieged in Khartoum by the Mahdi's fanatically brave Dervish followers, the 'fuzzy-wuzzies' of contemporary military legend. Gladstone was then persuaded to despatch a relief expedition, its commander General Lord Wolseley, the famous Victorian soldier-hero. But Gordon was speared to death in January 1885 as Wolseley's columns were on the verge of reaching Khartoum, after a long journey down the Nile, the country's sole means of communication between north and south. The British Government thereupon ordered Wolseley to pull back to the Egyptian frontier and abandon all plans for overthrowing the Mahdi, a Russian incursion into Afghanistan having brought about an international crisis and the call-up of the British reserves.

The 1st Green Howards arrived in Egypt in the August of 1884, but instead of joining Wolseley's relief expedition they were fated to spend the next nine months on uncomfortable garrison duties, although two officers and sixty men did see some action as members of a mounted infantry battalion in a subsidiary campaign around Suakin on the Red Sea. Not until March 1885 did the rest of the Battalion embark in four barges and two steamers to travel up the Nile to Assouan. There it spent a sweltering summer watching the frontier until, at last, it was ordered forward as part of a two-brigade force whose task was to halt a major Sudanese incursion. At Wadi Halfa, reached on Christmas Day, orders were received to discard khaki uniforms and put on red serge tunics, blue trousers and putties. Thus arrayed they would appear more formidable to the Dervishes.[17]

Five days later, the two brigades moved into the attack against the Sudanese, entrenched in a bend of the Nile, the Green Howards escorting its brigade's mule battery and a machine-gun battery of Gardner's, manned by men of the Battalion. After clearing the loop-holed houses of Kosheh, their brigade's task was to converge upon the nearby village of Ginnis. The Dervish fire from Kosheh was brisk but inaccurate, while the combination of the guns and long-range volleys from the British Martini-Henry hammerless breech-loading rifles, sighted to 1,000 yards, was too much for them. Slowly the Sudanese withdrew across the plain under this intense fire, their mounted emirs arousing the admiration of the Green Howard soldiers as they coolly rode among their men. The action had cost the British and Egyptian units forty-five officers and men killed and wounded. The small victory had restored the prestige of British arms and, for the time being, had halted any attempt by the Sudanese to invade Egypt. It had been an odd mixture of the old and the new, the last time the British Army wore scarlet in action and the first time the Green Howards had watched the effect of the new machine-guns.

Not until 1898 did the British settle this Sudanese threat to Egypt's stability once and for all, a task that was efficiently accomplished by an expedition commanded by General Sir Herbert Kichener. The decade and a

Above: The engagement at Ginnis in 1885. It was the last time the British Army wore scarlet in action.

half of rule of the Mahdi and his successors had been marked by religous intolerence, cruelty and slavery, the consequences of which had led to the death of some two-thirds of the country's population. The subsequent half-century of British rule gave the country a brief taste of peace, justice and comparative prosperity, justifying perhaps the death (mostly from disease) of many young British soldiers in that desert country.

The claim for a Battle Honour for Ginnis, made in 1937, was rejected in the standard arbitrary manner. The government had at the time refused to grant a medal for this small but strategically successful action on the grounds that the fighting had been insufficiently severe.[18] Treasury stringency has, over the years, been a primary factor in the notorious parsimony with which medals and decorations have been awarded to British servicemen. In claiming the Battle Honour, the point was to be made that 'the number of British corpses has not generally been a principle' in their award.

Modest indeed the battle losses had been, but soon afterwards, when the force returned to Assouan, the incessant deaths from enteric and heat-stroke so damaged the morale of the units stationed there that music and volley firing over the graves was forbidden at funerals. Before moving to Cyprus in 1887 1st Battalion was to lose eighty-four officers and soldiers from disease.

Cyprus at this time was nominally part of the Ottoman Empire, but since 1878 the British had administered and garrisoned the island, the result of a defensive alliance between Britain and Turkey intended to curb Russian expansion into Asia Minor. A summer spent in Troodos, the mountain hill-resort, did much to restore all ranks of the Battalion to good health, their fitness and freedom from disease being specially mentioned by the GOC when they left the island two years later; the officers, so he wrote, had 'supported sports and healthy recreations for the men',[19] a side of soldiering at last being given official encouragement.

Landing at Portsmouth in September 1889, the 1st Battalion was at hand to bid a festive farewell to the 2nd, which sailed for India on 1 January 1890 after twelve years in England and Ireland (the Cardwell system was now in full swing). It was the first time the two Battalions had ever met. New Year's Eve was an appropriate evening for the officers of the 1st to dine those of the 2nd, together with a number of retired and extra-regimentally employed officers, among them four generals amd three full colonels.

***

For the 2nd Battalion it was to be the start of a nineteen-year tour overseas. In October 1892, after nearly three years at Bangalore in southern India, it moved to Upper Burma with the prospect of active service. Ahead of the main body had sailed three small parties for conversion to mounted infantry platoons, similar to those raised in Egypt from the 1st Battalion for service in the Sudan. It was another instance of the need of mobile reconnaissance and patrolling troops in the Army: British regular cavalry regiments were at the time untrained for the role, their purpose being no more than an *arme blanche* – a heavy assault weapon for use in the shock role. Countrymen used to horses were plentiful in the Green Howards, so mounted work on the small local ponies came naturally to them.

Since the Green Howards had last served in Burma twenty-four years before, the whole country had come under British rule. Upper Burma had been annexed after the Third Burma War in 1885, a conflict that had arisen because of the threat of French penetration from their colonies (what is now Laos) on Burma's eastern border; nor could what was a grossly misruled region be allowed to exist on India's eastern frontier. The result was that once again the Green Howards were stationed on Burma's border, but this time several hundred miles to the north; for most of their tour they were to be based upon Schwebo and Bhamo.

This Upper Burma was as yet in a far from settled state. The many bands of marauding dacoits were able to find refuge across the Chinese border, and the then primitive Kachin tribesmen were reluctant to accept outside interference. Nevertheless life for the soldier was not usually too unpleasant. From time to time routine was disturbed by the excitement of a punitive column; half a century or so later, during the Second World War, British soldiers, their enemy the Japanese, sweated across those same jungle-covered Kachin Hills, insect bitten and fever-racked. They were not the first to do so. Green Howards during 1893 knew similar misery, as their columns chased an elusive enemy.

An account of one of these small expeditions, which returned to Bhamo in March 1893,[20] is noteworthy for being the earliest report of a Battalion's experiences on active service to appear in the new regimental journal, the first edition of which was delivered from the printers only on 18 April of that year. For the first four years its title was *Ours*, the change to *The Green Howards' Gazette* being made in April 1897; the subscription was 3 shillings annually for officers, half that sum for other ranks and their friends. The most prominent among the small body of the 1st Battalion enthusiasts responsible for the project had been Major J.W.R. Parker, its first editor and the Regiment's first serious historian. Captain M.L. Ferrar who succeeded him three years later was to keep the chair until 1942. His collection of militaria formed the basis of the Regimental Museum; his magnificent contribution to the Regiment's history is recorded in the bibliography. One of the objects of *Ours* had been to collect information which would help in compiling a history of the Regiment. In this it succeeded. For many years there was no lack of copy: even Crimean veterans were still around to set down their experiences. Without the regimental journal it would have been impossible for either Ferrar or his successors adequately to record the Regiment's achievements.[21]

★★★

A much more serious small war awaited the 2nd Battalion when it returned to India in early 1896. In the summer of 1897 mobilization orders for active service on the frontier were received, shortly after the Battalion

Below: A Green Howard mounted infantryman of the 1890s in Mandalay, Burma.

had celebrated at Raniket the Diamond Jubilee of the Queen. There had been a week of festivities, including a ball, a gymkhana and theatricals by 'The Green Howards' Dramatic Club'; huge bonfires on the surrounding hills had formed part of a vast chain fired simultaneously at every hill station along India's northern border. It was unfortunate that this celebration of imperial pride should have coincided with the most formidable rising ever of the Pathan tribesmen of the North-Western Frontier. Stirred to religious fervour by their Mullahs, the border between India and Afghanistan was quickly ablaze. Isolated forts and garrisons were besieged or attacked; a number of serious setbacks were suffered. The security of the entire Frontier was in jeopardy.

The most powerful and numerous tribe were the Afridi who, with the Orakzai, occupied the area known as the Tirah, to the west of Peshawar; about the size of Essex, its rugged mountains were bisected by fertile and prosperous valleys. Little of this unmapped Tirah had previously even been seen by any European. The boast of its warlike inhabitants, opponents far more formidable than the Black Mountain tribesmen whom the Regiment had fought a quarter of a century before, was that neither Mogul, Afghan, Persian or British army had ever penetrated their country.

The 2nd Green Howards was commanded by Lieutenant-Colonel W.E. Franklyn, the holder of a Staff College certificate, only the fourth Green Howard to be so qualified; the founder of a famous Regimental dynasty, both he and his son were to become generals and colonels of the Green Howards; his grandson was to be killed with the Regiment in the Second World War. The 17,000-strong force which his Battalion joined at Kohat, poised to penetrate the Tirah, consisted of one-third British troops and the rest Indian. But the total number of human beings assembled for the expedition far exceeded this figure. Because religion and caste barred Indian troops from menial tasks, a long tail of civilian 'followers' had always to accompany any force. Their number in this case exactly equalled that of the combatant troops; transport animals – mules, bullocks, pack-ponies and camels – came to no fewer than 42,810.

The Tirah Expeditionary Force was divided into two divisions: the 2nd Battalion of the 'Yorkshires', as they were generally known, being part of 2 Brigade of 1st Division. On 18 October these Green Howards took the dusty road out of Kohat, their division being the second to move off. The problem as ever on the Frontier was to protect the snake-like columns of fighting troops, guns, transport and followers, often confined to a single mountainous track. Movement was possible only if the hills were held by platoon-sized picquets, pushed out on either side as the column made its tardy progress forward and withdrawn as its tail passed through. Mountain artillery and the few Maxim machine-guns covered the picquets as they clambered up and ran down-hill, but the Pathans, many of them discharged soldiers of the Indian Army , were often well armed with captured, stolen, bought or home-made Lee-Metford magazine rifles. The speed and aggression with which the tribesmen followed up the withdrawing picquets could be dismaying; their shooting was effective. Every march could become a running battle, a succession of small withdrawals in which no wounded could ever be left behind to face a horrible death, often at the hands of the women.

The Regiment's part in this Tirah campaign is well documented, the colonel, the adjutant, two company commanders and an orderly corporal having all recorded their experiences.[22] The briefest and best was that of the adjutant, Lieutenant Ronald Fife, who wrote:

> Our first experience on entering Tirah was distinctly unpleasant, for, during daylight, we were attacked from all sides by a foe we seldom saw; and having bivouacked in a rather exposed place, we had a number of casualties. When darkness fell the enemy tried to press the attack home, but the light of star shells, fired by the artillery, enabled us to repel them. Next day we moved to attack the Sampagha Pass. Covered by the fire of some mountain batteries, we toiled uphill for several hours, during which the enemy never showed themselves but kept up a brisk fire from behind rocks. A great friend of mine, an artillery officer, was shot through the heart while I was talking to him . . .[23]

It was the first time the 2nd Battalion had seen action as a complete unit. Lance-Corporal Jones described the eagerness of everyone to get into action, and their reactions when the bullets started to fly. The next day Fife was slightly wounded during the assault on a similar pass, the descent from which:

> brought us right into the heart of the Tirah, where we spent a very cold and uncomfortable night, as the ration mules had not arrived and we had nothing to eat. We remained some time in this camp, where the enemy shot at us both by day and by night. When foraging or reconnaissance parties went out, there was sure to be fighting on the way back to camp, as the enemy always followed; and a sad disaster overtook a company of the Northamptonshire Regiment, which was almost destroyed when only a mile or so from camp.[24]

Fife remarked upon the intense night cold for men dressed in thin khaki. Although before leaving Kohat the Battalion had been issued with thick khaki-coloured peajackets, Cardigan waistcoats, Balaclava caps and warm mitts, this thick clothing was carried with the men's kit loaded on to mules during the march so was not always available when needed – the temperature could vary a hundred degrees Fahrenheit between dawn and midday.

It so happened that the first European to see the Tirah Plain was the leading Green Howard company commander, Lieutenant N.V. Edwards (or 'Nutty' as he was everywhere known), of whom we shall hear more later. This fertile country was now to be devastated by the invaders, the only method by which the Pathans could be driven to submit. Houses were burnt and razed, food stores destroyed and fruit trees ringed. Other armies might have slaughtered any inhabitants they could catch, but the British Army did not practise what is now known as genocide.

The Green Howards formed part of several columns that in turn penetrated even farther into the Tirah interior, but in early December the 1st Division began its long and arduous return journey, every Afridi valley having been well laid waste. It was none too soon, the first snow having already fallen. Not until 19 December did the Battalion pitch camp at Jamrud, just a mile from the entrance to the famous Khyber Pass. The men were to have little rest. On Christmas Day a four-day foray was made up the Khyber at a cost of four casualties, and a month later three companies undertook yet another but far longer expedition.

During their withdrawal from the Tirah the 1st Division had met little resistance, but the 2nd, marching by a different route, had to fight desperately for five days before it could extricate itself from those inhospitable hills, bearing its wounded on *dhoolies* – stretchers slung on poles carried by four men. During the retreat 200 more men were hit, the division barely avoiding complete disaster. In all the Tirah Expeditionary Force had lost 1,150 killed and wounded; many more men had been evacuated sick. No one appears to have counted the number of unfortunate followers who failed to return.

One Green Howard officer and nine other ranks had been killed; three officers and twenty-nine other ranks had been wounded. One of the wounded officers, Lieutenant O.C.S. Watson, had been shot through the lung while recovering a wounded lance-corporal and the dead body of his company commander, the latter to avoid its inevitable mutilation. Few then survived such an injury, but Watson did so only to earn a posthumous Victoria Cross in the First World War.

The Afridi had had enough. Clan after clan submitted after realizing the futility of further resistance. For a time an uneasy peace again held sway along the Frontier. But it had been an expensive little war.

★★★

The roll of Green Howard battle casualties did not tell the whole story of the Tirah campaign. Thirty-four more men died from other causes, most of them from disease; and, when the Battalion moved down to Peshawar in June 1898, even more were to succumb to the after effects of that bleak campaign.

The 2nd Battalion was to serve in India for eight more years. With little to break the monotony of life, one well remembered and glittering occasion was to be the Great Durbar of 1903. In this the Green Howards formed part of the 29,616-strong parade at Delhi where King Edward VII was proclaimed King-Emperor amid all that oriental pomp and colour unique to the old Indian Empire.

Over the years officers and men came and went from the Battalion – among them later in 1903 a detachment of two officers and sixty-five men under 'Nutty' Edwards as mounted infantry for British Somaliland, another harsh land of bare rocks. There one of a succession of campaigns was being fought against the so-called 'Mad Mullah' who, for more than twenty years, had been raiding tribes friendly to the British, always retreating over the border into Ethiopia (or Abyssinia as it then was) if hotly pursued. Hard marching, first on ponies and then on foot, on a gallon a day of filthy, sulphurous water was to be the lot of the mounted infantrymen. Blazing sun during the day alternated with bitterly cold nights; saddle blankets were their bedding. Yet among those Green Howards there was practically no sickness, nor did the Somalis succeed in finding a target among them.[25]

When, after two years of routine garrison duty in a South Africa that was making a rapid recovery from the Boer War, this 2nd Battalion at last arrived home on 1 March 1909, only the colonel, the quartermaster and half-a-dozen other ranks remained of the men who had sailed with it to India nineteen years earlier. During those years, 3,379 Green Howards had served in its ranks, only 223 of whom had died from various causes, a comparatively small number and a clear indication of improved living conditions and medical care.

To look ahead a little. Much that was to be taken for granted by later Green Howards was then evolving. Soon after the Battalion returned, the Green Howards' Association was founded on the Alma Day of 1909, an occasion which appears to have been first celebrated at the Curragh by the 1st Battalion in 1896.[26] The Association's rules then drawn up stood the test of time almost unchanged. In that year also, Fulford Barracks in York, the 2nd Battalion's station, saw the first Association Dinner.

The Officers' Dinner was already a long-established event, the first recorded being held at Klun's Hotel in Covent Garden in 1856. Until 1904 the dinner moved around a number of long since forgotten taverns in the City, Trafalgar Square and Regent Street, one especially popular hotel being the 'Ship and Turtle' in Leadenhall Street. The Cafe Royal and the Ritz were used in the first two decades of the century, after which the dinner settled into one of the service clubs, at first the Army and Navy, known to all as the 'Rag', and then the United Services – the 'Senior'.[27] Not until 1968 did it finally move to Yorkshire, convenient to the Territorials, by then and not before time, wholeheartedly accepted as a part of the Regiment.

***

We left the 1st Battalion at Portsmouth in 1890, saying farewell to the 2nd as it left for India. That July the even tenor of life was sharply disturbed. A battalion of Guards on London duties refused to turn out on parade – the Metropolitan Police were said also to be disaffected at the time. The result was that the Green Howards were rushed from Portsmouth to Wellington Barracks.[28] Their stay was short, but the severity of the crisis is perhaps indicated by their being inspected, when they entrained at Waterloo, by no less a person than the Princess of Wales herself, accompanied by her husband, the future King Edward VII, the old Duke of Cambridge as Commander-in-Chief, Lord Wolseley and the entire staff at the Horse Guards.

During the next decade the 1st Battalion was switched from England to Jersey to Ireland, and then in 1898 to Gibraltar. During its year there it was announced, to much rejoicing, that a more understanding Secretary-of-State for War had at last restored to the Regiment its green facings, so wantonly removed seventeen years before.

Back in England in 1899, orders were received in November to mobilize for the Second Boer War. As the Green Howards marched out of their barracks in Bradford, the centre of the city was crammed with a mass of humanity, estimated at some 50,000, cheering their heads off and breaking into 'God Save the Queen' as the troop-train steamed away. It was a measure of the patriotic fervour for this war that was sweeping the country.

Nearly a century later, the rights and wrongs of the Boer War are still being argued. Here they can be no more than touched upon. Ever since Britain's acquisition of the Cape of Good Hope during the Napoleonic Wars, the Dutch farmers already settled there had been trekking ever farther into the interior in an endeavour to retain their independent way of life, and always at odds with the African tribes whose land they seized. Behind them lay the two British colonies of the Cape and Natal, part of an increasingly expansionist late-Victorian Empire. From there acquisitive eyes were being cast northwards, not only towards the new Boer republics of the Orange Free State and Transvaal, but also to the vast territories beyond, especially the land that first became Rhodesia and is now Zimbabwe. During the First Boer War fought in 1881, the Boer irregular horsemen, accustomed from boyhood to the saddle and the rifle, had inflicted a disastrous and unforgiven defeat upon a small force of British regulars at Majuba Hill. The discovery of diamonds and gold in the two republics produced further tensions: immigrants to the Transvaal, most of them British and lured there by the gold, were increasingly being persecuted.

Rich now from their gold and diamonds, the Boers were able to buy modern weapons to arm their 50,000 mounted infantrymen. Both sides now felt threatened and when the British began to reinforce their small garrison, the Transvaal pre-empted them. On 9 October the Boers' President, Paul Kruger, despatched an ultimatum demanding, among other things, that British troops already at sea be turned back. Rejection was inevitable. Two days later the Boers invaded north Natal.

The 1st Battalion landed at Cape Town on 15 December 1899 at the end of the 'Black Week', so called because three separate British forces, advancing northwards along the lines of the railways, had all suffered serious reverses. Losses among officers, picked off by the Boer marksmen, had been heavy, and one of the first orders received by the Green Howards was for officers to dress the same as their men, for swords to be placed in store, buttons dirtied and badges of rank removed. The instructions caused surprise and some resentment. Such lessons are quickly forgotten however; swords were again carried into action in 1914 and officers' uniforms were again very different from those of other ranks.

Entrained northwards to the borders of the Orange Free State, the men, half of them unfit reservists, were quickly toughened, learning to carry fifty-pound loads over long distances in the grinding summer heat. The accuracy of the Boer riflemen was first felt when, in the Colesberg area, a half-company of Green Howards, with the same number of New Zealanders, were holding one of those steep, bare *kopjes* (later to be known as New Zealand Hill) that spring suddenly out of southern Africa's far-stretching grass-covered veldt. Attacked by an overwhelming force of Boers on 15 January, the position was held in an unpleasantly costly struggle during which the company commander was seriously wounded and his colour-sergeant killed. At the turn of the year Field Marshal Lord Roberts, whose reputation had

been made in India, had arrived in South Africa, despatched after 'Black
Week' to supersede General Sir Redvers Buller. Ladysmith in Natal,
Mafeking on the north-western border of the Transvaal, and Kimberley,
centre of the Orange Free State diamond industry, were then all under close
siege. The Boer General Cronje, blocking the railway line leading to
Kimberley, had already inflicted two disastrous reverses on British columns.
Roberts decided, therefore, to concentrate his strength in this area, defeat

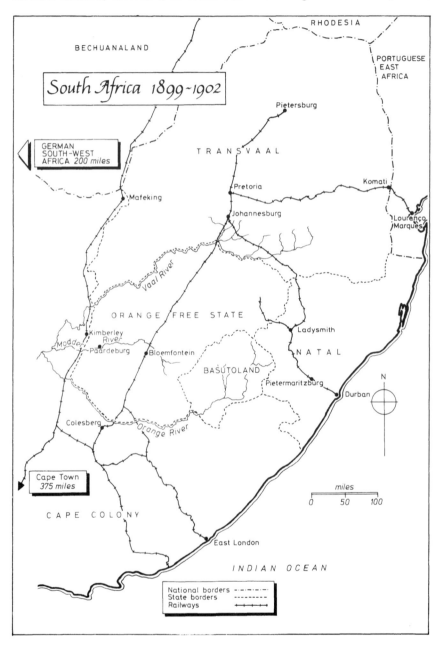

South Africa 1899-1902

Cronje, relieve Kimberley and thrust first towards Bloemfontein, the Free State capital and then onwards to Pretoria.

Quietly switching some of his troops westwards, the Green Howards among them, Roberts with 40,000 men boldly cut loose from the railway, directing his cavalry division towards Kimberley and marching his infantry across Cronje's flank towards Bloemfontein, 100 miles away across the veldt. Boer generals were as capable of error as the British, and Cronje, reluctant to lose his valuable ox-wagons, tied himself to their slow pace, thereby sacrificing his horsemen's mobility. At Paardeberg on 18 February he was trapped with his force entrenched north of the River Modder. Other Boer forces were nearby and Kitchener, temporarily in charge because of Roberts's sickness, decided upon an immediate attack against Cronje's laager so as to forestall his relief or escape.

The Green Howards had left the railway line six days earlier,[29] the start of a ten-month-long march, during which they slept under cover for only four nights. Freezing bivouacs alternated with tropical-like downpours. Even before they left the railway, uniforms were in rags and boots almost worn out. Five large biscuits and a pound of tough trek-ox was the basis of the daily ration, bully beef a luxury.

Needless to say when they halted at a prosperous farm on 17 February, its bounty of goats, pigs, chickens, ducks, peaches, apricots, grapes, flour and groceries was grasped with delight. Pushed on from the looted farm towards Paardeberg that night, the weary and soon very thirsty men were quickly thrust into the attack. As well as clearing the Boers from the south bank of the Modder, the task of the brigade to which they were temporarily attached was to distract the enemy while a pincer movement crushed the Boer main position on the north bank of the river.

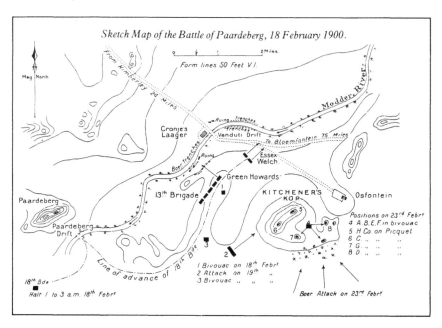

Sketch Map of the Battle of Paardeberg, 18 February 1900.

Fifteen hundred yards of featureless and forward-sloping veldt lay between the Green Howards and the river. Soon the characteristic 'pick-pock' of the Boer Mausers grew in volume; bullets phutted angrily into the earth and men began to tumble over, easy targets for the Boer marksmen. Advancing by short sectional rushes, only some sixty men managed to reach the water's edge. There they found the south bank clear of Boers. That murderous fire had come from well-entrenched riflemen on the far side of the unfordable thirty-yard-wide river. Some Green Howards found shelter of a sort near the bank, but most lay out under fire on the almost bare veldt until nightfall. During the course of the day several men, with superb courage, crept forward to collect water for the thirst-maddened wounded. Among them was Sergeant Atkinson, who seven times made the journey before being mortally wounded; his subsequent Victoria Cross was one of the first to be awarded posthumously and the third to be won by the Regiment.

In the end the Boers are said to have held their fire out of pity for their enemies, a gesture unrecorded in Regimental accounts of the battle.[30] But the cost had been heavy. The veldt was dotted with dead, dying and wounded men, thirst-stricken and tormented by the sun. In all, forty-one Green Howards had died and eighty-five had been wounded. At no time during the day had anyone seen a Boer. For the first time the Regiment had felt the power of concealed riflemen, using smokeless ammunition. The main attacks elsewhere had been sadly ill organized, and although the day ended with Cronje's laager in flames, pounded by the British artillery, and most of his horses and oxen dead, it was another nine days before he surrendered. Out of that stinking knacker's yard of burnt wagons and rotting animals, 4,000 Boers marched into captivity. Kitchener had deployed some 16,000 men.

Below: The Boer General Cronje surrendering to Lord Roberts after the Battle of Paardeberg. The saddle on his emaciated horse is on show in the Regimental Museum.

Not only had Kitchener mounted the attack with too little preparation, but he had then lost all control. Untrained and unpractised as a staff officer in the field, like Roberts he had no concept how a staff should be used: incapable of decentralizing, he tried to do everything himself. Nor had he ever served with troops as a young officer, or even commanded a unit or formation of British soldiers. A fine politician and diplomat, other than dash and determination he lacked nearly every quality required of a general. Although Paardeberg shone as a victory after a successsion of defeats, it was probably the worst managed battle of the Boer War, notable only for the magnificent courage, discipline and endurance of units like the Green Howards, who, in the words of the Official History, 'had very thoroughly carried out, if not exceeded their task'.[31]

Before Cronje surrendered, the Regiment had twice again been in action. As dusk fell on that first day, General Christiaan de Wet with only 300 men had, in a superbly bold stroke, seized an isolated and commanding *kopje* about two miles south of the river, afterwards known as Kitchener's Kop. In the late afternoon of 19 February the Green Howards took part in a brigade attack that failed to drive de Wet off the feature. There the Boers clung for three days, a major aggravation for Roberts, now again in command, and one that all but forced him to abandon the battle. No sooner had de Wet been forced off the *kopje*, however, than he met with reinforcements, so persuading him to attempt its recapture. By then the Green Howards were defending it. At dawn on 23 February the Boers gained a lodgment on the hill's lower slopes but the subsequent fire fight, continuing for several hours, took a heavy toll of their horses. In the end they withdrew, leaving eighty-five of their number as prisoners of the Green Howards. Among their dead was counted one man of more than seventy years; others were mere boys. The battle had cost the Regiment another eleven dead and twenty wounded.

The fate of the wounded was wretched. There was a gross shortage of medical supplies, including chloroform for the surgery, and few doctors or orderlies. Only unsprung ox-wagons were available for the long journey over the boulder-strewn veldt to the railway; one wounded NCO afterwards wondered why Dante had never described that particular torment.[32]

★★★

After Cronje's surrender the Green Howards remained on the exposed summit of Kitchener's Kop, for much of the time in pouring rain, their rations a couple of biscuits and half-a-pound of stringy meat daily. Tobacco was non-existent. But when Roberts set out for Bloemfontein on 7 March they were still fit enough to march for a week with little water, and fight an action at Driefontein on the way, one that cost them a further twenty-eight killed and wounded.

When the Battalion at last made its ceremonial entry in rags and broken boots into the captured Free State capital, the more disreputable-looking men were hidden in the inner ranks of the column. About twenty more men died in Bloemfontein from the ubiquitous enteric fever and dysentery, but

*Above: An improvised Officers' Mess of the 3rd Militia Battalion
near Bloemfontein.*

losses in other regiments were far more severe. During the pause for supplies
to be brought forward over the long line of communications, the Green
Howards enjoyed a little rest, but spent far more of their time following up
often nebulous reports of Boer parties nearby. Among the reinforcements
received at this time was a company drawn from the Regiment's Volunteer
battalions, described as a fine body of men. They were a part of that surge
of civilians, not only from the British Isles, but from Australia, Canada, New
Zealand and South Africa as well, who volunteered for the war. It was an
astounding display of loyalty to the 'mother' country, to be twice repeated
during the next half-century.

During the subsequent advance to Pretoria, the Transvaal capital, these
Yorkshire Regiment Volunteers tasted action when the Boers tried to delay
Roberts's 40,000-strong army, but they suffered no casualties. Then, after
halting briefly at Johannesburg, founded only fifteen years earlier, the Green
Howards finally entered Pretoria on 5 June.

Since Driefontein, the Battalion itself had lost only one man wounded,
but the 2nd and 4th Battalions of Mounted Infantry, the first containing a
section and the second a company of Green Howards, had seen far more
excitement, skirmishing ahead and on the flanks of the marching columns,
and suffering several casualties in the process.

After a pause of some weeks in the Pretoria area the Green Howards
formed part of the force that continued the chase eastwards to the
neighbourhood of Komati Poort on the borders of Mozambique, where they
arrived in mid-September. They had walked a thousand miles. President
Kruger had fled to Europe and the war seemed to be over.

But it was not. The Boers had been beaten in the field by an army that was soon to number nearly half a million men. Under the inspired leadership of de Wet, however, the Boers turned to the guerrilla warfare for which they were so well suited, sabotaging railways and bridges, ambushing supply columns and attacking isolated posts. When Roberts left for home in November 1900 Kitchener took over, faced with the problem of protecting the enormously long lines of communication and harrying the Boers to submission. It was to be another eighteen months before overwhelming numbers and a devastated countryside forced them to surrender.

During these wearisome months most of the infantry, including the 1st Battalion, spent much of its time manning small picquets or occupying corrugated iron and stone blockhouses, linked by barbed-wire and sited within rifle shot of one another. With these Kitchener protected the railway and enclosed vast sections of the veldt through which the mounted troops conducted drives, a very few successful, many not so. In these blockhouses Green Howards baked and sometimes froze in section-sized detachments, only rarely hearing a shot fired. It was now a light cavalry, not an infantry, war and the Green Howard Mounted Infantry on their hardy ponies were always busy, the envy of their regimental comrades, condemned to sweltering boredom. As Kipling wrote:

I wish my mother could see me now, with a fence-post under my arm,
And a knife and spoon in my putties that I found on a Boer farm,
Atop of a sore-backed Argentine, with a thirst that you couldn't buy,
I used to be in the Yorkshires once . . .
But now I am M.I.[33]

This low-lying border area where the Battalion was based until the following August was especially unhealthy, at one time ten officers and some 300 men being in hospital. A year earlier the Green Howards were, as one of their officers wrote, 'as fine a body of men as one could wish to see; perfect in health, and physically fit to go anywhere'. They left for Natal with 'every man saturated with malaria, and quite unable to do any severe work . . .'[34] From Natal the now under-strength unit moved back to the Johannesburg area where it remained, split into small detachments, until four months after the surrender of the last Boers in the field in May 1902.

By dogged patience, Kitchener had won a guerrilla war. By today's standards, he had done so quickly. What is more, there had been little brutality. Much propaganda has been made of the death toll from disease of Dutch women and children, interned in the so-called 'concentration camps'. The cause, however, was gross administrative inefficiency. At the time many Boers, out on commando and fearful for their families alone on their farms, had welcomed their being concentrated in this way.

It was often suggested that other European powers, faced with the problem, might have ended the Boer War more quickly by using rather different methods. But to have done so would have produced lasting bitterness. As it was, a Union of South Africa was granted dominion status only eight years later, and Boers in large numbers were to volunteer to fight alongside their recent bitter enemies in both the First and the Second World Wars.

As well as 1st Battalion and the Volunteers, the Militia too served in South Africa, although contemporary regimental histories fail to mention the fact.[35] In the army reorganization of 1881 the North York Militia had become the 4th Battalion of the Regiment with the Knaresborough Militia, a quite modern unit, taking its place as the 3rd Battalion, much to the disgust of that ancient North Riding unit. This 3rd Battalion was embodied in December 1899, the 4th Battalion the following May. For the greater part of the war the latter served at home, finding drafts for both the 1st Battalion and the Mounted Infantry, and lining the streets for the funeral of Queen Victoria on 2 February 1901. At last, in March 1902, the unit sailed for South Africa to spend a few months on blockhouse duty before peace was declared, its only casualty being an officer shot accidentally by a sentry, that far from infrequent tragedy. The 3rd Battalion, on the other hand, spent most of the war in South Africa, but no record of its service there is known to have survived.

The company of volunteers which reinforced the regulars had come from the the 1st and 2nd Volunteer Battalions of the Regiment, based at Northallerton and Scarborough. When the men arrived home eighteen months later, having earned the respect of their regular comrades, the North Riding feted them. At a welcoming breakfast in Richmond Town Hall they were reminded that they had been welcomed as Green Howards; for the first time a line battalion had been helped by its Militia and Volunteers and they could now really claim to belong to the Regiment.[36] It was something that had not been said before. Nevertheless, the amateur soldier had to wait until the First World War, if not the Second, before he earned the recognition that was his due.

Below: An armoured train somewhere in the veldt manned by the 3rd Militia Battalion.

Above: Captain C. A. C. King's charger protected in a dugout from both the sun and enemy fire.

Below: A block-house at Rhenoster Bridge.

The Eastern
Mediterranean
and the Middle East

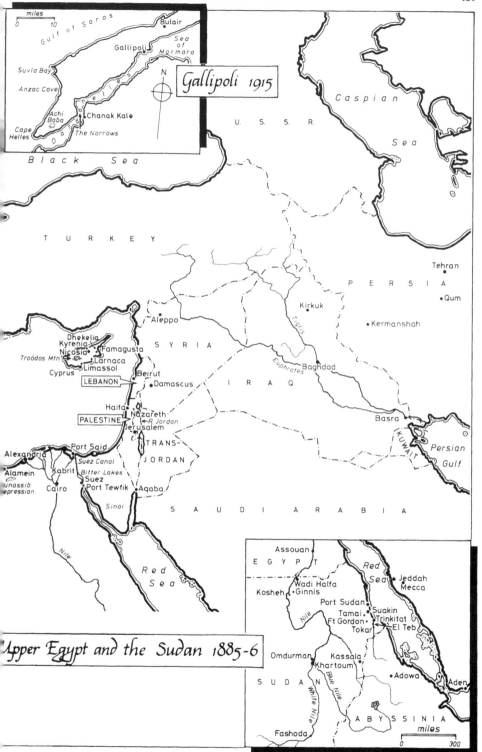

Gallipoli 1915 (inset map)

miles
0   10

Gulf of Saros
Bulair
Gallipoli
Sea of Marmara
Suvla Bay
Anzac Cove
Achi Baba
Chanak Kale
Cape Helles
The Narrows
Black Sea

N

U. S. S. R.

Caspian Sea

T U R K E Y

P E R S I A

Tehran
•Qum

Aleppo
Kirkuk
•Kermanshah

S Y R I A
Tigris
Euphrates
Baghdad

Dhekelia
Kyrenia
Nicosia• •Famagusta
Troödos Mtn •Larnaca
Cyprus •Limassol
Beirut
•Damascus
LEBANON

I R A Q

Basra

Haifa
PALESTINE •Nazareth
←R Jordan
Jerusalem

TRANS-
JORDAN

Port Said
Alexandria
Suez Canal
Alamein Kabrit Bitter Lakes
unassib Suez
epression Cairo Port Tewfik
•Aqaba
Sinai

KUWAIT

Persian Gulf

S A U D I   A R A B I A

Nile

Red Sea

Upper Egypt and the Sudan 1885-6

Assouan
E G Y P T
Red Sea
•Jeddah
Mecca
Wadi Halfa
Kosheh •Ginnis
Port Sudan•
Tamai• •Suakin
Ft Gordon• •Trinkitat
Tokar El Teb
Omdurman Kassala
Khartoum
•Adowa
Aden
S U D A N
White Nile
Blue Nile
Fashoda
A B Y S S I N I A

miles
0   300

Left: An Edwardian luncheon party held in Major C. A. C. King's bungalow at Wynberg, Cape Colony, on the King's Birthday, 9 November 1907. He is sitting next to the uniformed naval officer. Behind him is Colonel E. S. Bulfin, a future Colonel of the Regiment.

Below: The Officers' Mess of the 2nd Battalion at Wynberg, Cape Colony, in 1907.

6. THE FIRST WORLD WAR AND ITS PRELUDE: 1902-15

The reverses of the 'Black Week', followed by the long-drawn-out delay in finishing off the Boers, had brought home to the British public and its politicians the unpalatable fact that its Army had been found wanting. At the same time an envious and increasingly hostile Germany was undertaking a vast programme of naval expansion, a challenge to Britain's maritime supremacy upon which depended both the country's security from invasion and the protection of those vital trade routes of an Empire that now covered a quarter of the world's land surface. And Britain stood isolated: the war in South Africa had aroused not only world-wide condemnation but much smug satisfaction at her poor showing. Of likely allies against Germany, there were as yet none.

The consequence was a drive for military reform. The lessons learned from the war were minutely dissected. Very soon a succession of measures followed one upon another with bewildering rapidity. The Committee of Imperial Defence, now to be headed by the Prime Minister, was given teeth. The appointment of Commander-in-Chief was abolished and the control of the Army vested in an Army Council under the Secretary-of-State for War. The War Office was reorganized with many of its over-centralized functions hived off to new command headquarters, and a General Staff was created. Salisbury Plain was acquired as a training area and new artillery weapons were ordered.

It was to be some time, however, before these reforms touched the daily life of the 1st Yorkshires, brought back from South Africa to Sheffield in September 1902. Khaki became the universal dress, the red tunic retained only for ceremonial and walking out purposes. A shortened rifle was issued, the previous weapon having proved too long and clumsy in battle; that famous new Short Lee-Enfield was to remain in use throughout two world wars and beyond. Otherwise little changed. Drill, a lot of musketry, ceremonial duties, guards and fatigues, with a little training in the summer, filled the soldiers' year.[1] As for the officers, for whom a private income was all but essential then and for the next forty years, their few hours of morning work were alleviated by generous spells of leave.

Despite their many virtues, few of these officers took a deep interest in their calling. Exceptions were the two Franklyns. The father, later Lieutenant-General Sir William Franklyn, who had commanded 2nd Battalion in the Tirah and been appointed Colonel of the Regiment in

October 1906, died soon after the outbreak of the coming war, having just achieved command of an army. His son, General Sir Harold Franklyn, who joined the 2nd Battalion in 1905 and in due course also became Colonel, was in old age unconventionally outspoken about his seniors and contemporaries. In an article entitled 'Regimental Personalities' written for *The Green Howards' Gazette*, he described how the 2nd Battalion went steadily downhill when a new commanding officer took over in 1906. The new colonel, although devoted to his regiment, had learned little about his profession during a long career; at fifty-three he found it difficult to begin. With such a commander, at a loss how to act, the Battalion was reported as being unfit for active service. Another officer, a man with quick intelligence and a magnetic personality who became a brigadier-general, might have risen to high command, Franklyn wrote, if only he had studied his profession as a young officer. One after another, Franklyn painted incisive portraits of brave, tough and colourful men, fine regimental officers when need be, but brought up by 'seniors who never looked beyond the regimental horizon' and whose outlook was narrow and devoid of new ideas.[2]

It had been thought that the abolition of purchase in 1871 would give the Army a more professional corps of officers, but the results had been disappointing. Purchase had at least produced some able young senior officers. After its abolition, promotion was governed by rigid seniority, alleviated only by a system of brevets, occasionally awarded for outstanding work, usually on active service. But these brevets pulled only the fortunate few out of the ruck. For the rest promotion was slow: until the start of the Second World War battalion commanders in peacetime could often be in their fifties and company commanders in their forties. Nevertheless, when the 1st Battalion moved in 1905 from Yorkshire to Aldershot, the Army's nerve-centre, it was well commanded and excelled in the training field.[3]

In 1906, during this Aldershot tour, Queen Alexandra presented new Colours to the 1st Battalion to replace those which, as a young Princess, she had confided to its care thirty-two years before when her connection with the Regiment first began. The ceremony took place, not on the parade ground, but in the drawing-room at Buckingham Palace. It was to be repeated three years later. In November 1909 the 2nd Battalion, having arrived home earlier that year after two years spent on garrison duty in South Africa, also received new Colours from her hands, but this time at Windsor Castle. But it was to be a further five years before the Regiment received Her Majesty as its first Colonel-in-Chief, the first royal lady to receive such an appointment – a 'notable innovation' as the *Daily Telegraph* described it.

***

The 2nd Battalion had replaced its sister unit on the Home Establishment, the 1st Battalion having left England for Egypt in January 1908 to begin another overseas tour. Before it left, the Army had undergone yet another series of major reforms.

At the end of 1905, Richard B. Haldane, a lawyer and philosopher with an incisively logical mind, was appointed Secretary-of-State for War in the new Liberal administration. When asked for his proposals at his first meeting with the Army Council, Haldane replied that he was no more than a young and blushing virgin, of whom nothing could be expected for nine months. The generals did not have to wait so long. Advised by a handful of officers (among them the young Major-General Douglas Haig), Haldane created from the forces at home a field army consisting of one cavalry and six infantry divisions, backed by proper administrative services. Adequate reserves and staffs were organized, staff training increased and expanded, and large-scale manoeuvres planned. And, at the same time, the Army Vote was reduced, the savings brought about by disbanding some unneeded units.

These regular formations were the front line. Behind them, earmarked for home defence and possible expansion, were the fourteen infantry divisions of the new Territorial Force, together with the same number of Yeomanry cavalry brigades, organized in a similar manner to their regular counterparts with the necessary supporting arms and administrative units. For the Regiment, the consequence was that its 1st and 2nd Volunteer Battalions became its 4th and 5th Battalions. Amateurs still they were, and inadequately equipped and trained, but they were greatly superior to the Volunteers they had replaced.

At the same time the 4th Militia Battalion of the Regiment was disbanded and the 3rd converted to one of the new Special Service units, its role to provide a reserve for the regulars in wartime and a source of recruits beforehand.

These changes to the part-time forces of the Crown inevitably produced the strong opposition to be expected among the many entrenched interests, overcome only by Haldane's flexibility and acumen. The Green Howards were especially angry because the 4th Battalion, successor to the ancient North York Militia, was to be lost, and the 'Johnny-come-lately' 3rd Battalion retained. The *Gazette* expressed 'regret and astonishment' at this 'vexatious disbandment'. A deputation of Green Howards to the War Office, led by H.M. Lieutenant for the North Riding, achieved nothing at all.[4]

Another subject was aggravating the Regiment at the time. In 1904 the Esher Committee had recommended without success that the old regimental numbers be restored, with the territorial titles retained as well. The proposal received only limited support, but some regiments, including the Yorkshires, were strongly in favour. It was not just sentiment. With five other regiments having 'Yorkshire' in their title, confusion was inevitable. In South Africa stores and mail had often been misdirected. The right to wear their old green facings having just been restored, surely the next step was to become again the 19th of Foot, a title which a commanding officer would sometimes alternate with 'Green Howards' when addressing his troops on parade. There is, however, no record of the Regiment having as yet summoned up the temerity to propose their much-loved nickname as its official designation.[5] It would have been an unusual innovation at the time.

★★★

When the 2nd Battalion first returned from South Africa in March 1909, it went to Fulford Barracks in York, with a two-company detachment at Scarborough. What was incorrectly thought to have been the Green Howards' first visit in their history to their own county was celebrated by the people of the North Riding presenting the Battalion with a fine silver centre-piece consisting of four soldiers in different uniforms grouped around the Colours and the Alma drums (at this time it had been forgotten that the Regiment had been stationed in York in 1750).

Under the new order, large-scale formation manoeuvres took place each summer. Fife, now a major, described those of 1912 as having 'began about mid-day and were prolonged until the afternoon of the following day without intermission, were rather exhausting. But although Sir Douglas Haig [now commanding at Aldershot], like everybody else, had been on the go all day and night, he always seemed perfectly fresh . . .'[6] By the standards of the day such training was, of course, intensive. It was about then that the old eight-company organization was replaced by four double-sized companies, a more flexible system that has, by and large, stood the test of time.

During the previous year Fife had been involved in more serious work. Training – in Wales that summer – had been interrupted by industrial violence in Liverpool and elsewhere, arising from widespread transport and dock strikes. When Lime Street Police Station in Liverpool was besieged and a large number of police injured, the Army was called out 'in aid of the civil power'. Half the Battalion, commanded by Fife, was deployed in the Scotland Road area, where a mixed mob of Orangemen and Nationalists, their differences sunk just for the once, was in control. Magistrates read the Riot Act and Green Howards dispersed the mob at the bayonet point. Fleeing into the adjacant houses, the rioters showered the troops with every sort of missile; large numbers of them had to be dragged out of the houses and handed over to the police. It was unpleasant work and, so far as is recorded, the first time since the early part of the previous century that the Regiment had been so involved against their own countrymen. Worse was to follow when troops of another unit had to open fire after being violently attacked. Two Irishmen were killed.[7]

***

There was a certain grim inevitability about the approach of the First World War – or the Great War as for years it was to be known. Britain's post-Boer War isolation had been short-lived. In 1904 began the first moves towards what became the *Entente Cordiale* between Britain and France. Both countries needed allies: France, anxious to recover Alsace and Lorraine, torn from her after her defeat in 1871, was at the same time fearful of an increasingly powerful Germany's ill-disguised ambitions. The *Entente Cordiale* eliminated colonial friction between Britain and France and resulted in the two nations settling outstanding disputes and making an effort to put aside mutual prejudices and live in some measure of harmony. In much the same way, Russia and Britain settled a few of their many differences in 1907, thereby all but enclosing the German Empire in a defensive ring that further increased the latter's fear of encirclement.

Germany's ally was the faltering Austro-Hungarian Empire, then encompassing much of the central Europe that was to fall under Soviet suzerainty forty years later. She was also Russia's bitter rival in the labyrinth of the Balkans. In Sarajevo on 28 June 1914, the murder by Serbian assassins of the Archduke Franz Ferdinand of Austria, heir to the throne of Austria-Hungary, together with his wife, was no more than the spark that ignited the time bomb that had been in place since before the start of the century. The major powers of Europe mobilized their vast conscript armies. Millions of men began to move towards their battle stations using the intricate railway systems constructed largely with that specific purpose in mind. In turn the great nations began to declare war one on another.

German strategy for a war on two fronts envisaged using the greater part of her strength to crush France, after which her forces would be switched eastwards against Russia. The Schlieffen Plan for this western campaign involved a strong right hook through Belgium, a move that risked bringing in Britain, as ever concerned with the independence of the Low Countries. On 4 August the Germans invaded Belgium. That evening the Foreign Secretary, Sir Edward Grey, looking out over Whitehall at dusk, observed, 'The lamps are going out all over Europe; we shall not see them lit again in our lifetime'.

Although no formal treaty existed between Britain and France, talks between the general staffs of the two countries had been in hand since 1906. At these it had been agreed that the British Expeditionary Force would be placed on the extreme French left, its role to help check the coming German offensive as the French armies thrust deep into Lorraine. This French attack was to fail, but meanwhile the British Expeditionary Force of six divisions of infantry, each numbering 18,000 men and nearly 6,000 horses, together with a single cavalry division, had crossed to France and met the full force of the German right hook at Mons. The French, together with this small British force battling on their left, were then pushed back through northern France until the Germans were at last halted on the River Marne as Paris seemed to be within their grasp. Then followed what was to be known as 'the race to the sea' with the opposing armies pushing ever northwards towards the Channel coast, striving to outflank each other.

***

The British mobilization had been masterly. In just five days 1,800 trains conveyed the first five divisions to the ports (two were delayed by Cabinet decision). Although 120,000 horses had to be bought or requisitioned, motor transport units, aircraft and even some armoured cars provided a glimpse of the future. As the official historian accurately stated, it was 'incomparably the best organized, best trained and best equipped British Army that ever went forth to war'.[8] Only in medium and heavy artillery, in machine-guns (only two in each battalion), and in its methods of co-operation between the infantry and artillery, was it inferior to either its allies or opponents.

But the 2nd Green Howards was absent from the original British Expeditionary Force. Stationed in Guernsey for the past year, it had not formed part of a field formation. Not until 28 August did it join the newly

Above: *Guard of Honour provided by the 2nd Battalion for the departure of the retiring Lieutenant-Governor of Guernsey, Sir E. D. F. Hamilton, on 7 April 1914.*

Below: *The 2nd Battalion marches off to the First World War on 28 August 1914 led by the Band of the Guernsey Militia.*

A Private Soldier 1742.

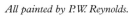

All painted by P.W. Reynolds.

B. Grenadier 1751.

C. Grenadier 1768.

D. Officer 1792.

E. Private Soldier 1832.

F. Bandsman and Drummer 1852.

G. Officer with Regimental Colour.

H. Officer and Private of Grenadiers 1854.

formed 7th Division taking shape in the New Forest. With its infantry consisting mostly of units withdrawn from overseas garrisons or Guards battalions taken off London duties, 7th Division lacked the cohesion of those originally earmarked for the BEF. However, far fewer reservists had been needed to bring units up to strength, and the soldiers were tough, mature men, averaging five years' service.

Busy days were spent during the last weeks of a lovely summer collecting war equipment from every corner of the country and hardening feet in endless training marches. News from France was sparse, but the Green Howards now stopped fretting that the war would pass them by: it had become clear that it would not be over by Christmas, but few envisaged the near holocaust that awaited them across the Channel.

At last, on 6 October, the first half of the 2nd Battalion landed at Zeebrugge and entrained for Bruges. As the first British troops to be seen there, they received a hearty welcome. The intention had been that their 7th Division, together with 3rd Cavalry Division, would help defend Antwerp, still holding out against the Germans, but news now arrived that the city was about to be evacuated. Plans were changed and the two divisions were instead directed eastwards towards Ypres, that lovely and as yet undamaged old city. Moving north towards them was what remained of the original BEF, much depleted after its retreat from Mons and the subsequent fighting on the Marne, and now playing its part in the attempt to envelop the seaward flank of the German armies. The first member of the Regiment to fire a shot at the enemy in the First World War was Lieutenant and Quartermaster E. Pickard, who had been awarded his DCM as a colour-sergeant on that sweltering river-bank at Paardeberg. With the advance party of the Battalion, he had been engaged on the mundane task of allotting billets in Ypres. Surprised by two lance-bearing Uhlans, he and his RQMS proved themselves the quicker on the draw. Certainly the first, it was probably the last time during the war that the Regiment shot at mounted German horsemen.

The next day, 15 October, 7th Division moved eastwards out of Ypres, its orders to attack in the direction of Menin. But faced by overwhelming numbers of German divisions, it was forced on to the defensive. After four days of weary marching and counter-marching, on the evening of 19 October the 2nd Battalion found itself a part of a thin defensive line along the forward ridges of the saucer of low hills overlooking Ypres – hills that for four long years were to be churned and churned again into a morass of corpses, mud and crushed buildings. As yet, however, dotted with woods and intersected by lovely little streams, the countryside had something about it of the Weald of Kent.

The task of the 1,000-strong 2nd Battalion was to hold a mile of front running eastwards from the crossroads at the hamlet of Petit Kruiseecke, one kilometre from Gheluvelt and nine from Ypres. Here or hereabouts it was to do battle for the next sixteen days. Fighting in small, isolated parties, it held what was little more than an outpost. But there was rarely anything behind those Green Howards. Picks and shovels were few and there was practically no wire. On most days mist enveloped everything until after midday.

Against the 7th Division when that First Battle of Ypres began were deployed two complete German army corps; moving up behind them were more enemy formations which would in turn be thrown into the battle. As the divisional historian admitted, it was difficult to get a clear picture of the confused fighting. War diaries provided only the bare bones of the story; the few participants who survived the war found it hard to recapture the details, even when they wished to do so.

Covered by heavy and well-directed artillery fire which the few British guns could do little to counter, on most days packed masses of Germans in solid ranks advanced time and again to be slaughtered by the well-controlled Green Howard musketry, each man capable of his fifteen aimed rounds each minute. To augment its riflemen, the Battalion possessed just two Vickers .303in machine-guns, but such was the intensity of the British fire that the enemy mistook the riflemen for machine-gunners. In any case, within three days the Battalion machine-gun detachment had been destroyed, every man hit and Lieutenant Ledgard, its commander, killed after a day spent heaving the heavy guns from position to position trying to avoid the enemy fire.[9] That day, the brigade war diary recorded 'The tenacity of this battalion was most remarkable; though subject to heavy artillery fire and constant attacks, it held on stoutly and never wavered'.[10]

Weary when the battle started, short of sleep, frozen at night and usually hungry, the Green Howards clung to their positions. Each night, 96,000 rounds of ammunition were brought up to the forward companies and then shot away by an ever-decreasing number of riflemen. In front of the rudimentary trenches scraped in the sandy soil, German corpses piled up in heaps. Many of them had been young volunteers, schoolboys and students, their junior leaders frequently schoolmasters, youths fanatically brave but with little training, recruited in August into reserve corps to augment the regular German conscript armies. Much of the flower of German youth died in Flanders that autumn.[11] But it was not just these volunteers with whom the Green Howards had to contend. Regular Bavarian units were engaged as well, among their NCOs a certain Corporal Adolf Hitler.[12]

It was doubly unfortunate for those young Germans that they were faced by the best infantry in the world – men who were not only long-service regulars, the only ones in Europe, but whose more senior officers and NCOs were alone in having fought against a European enemy and profited from the experience. The Boers had taught them the value of fieldcraft and sound markmanship. While the Germans, French and Russians still attacked shoulder to shoulder in solid ranks, the British regulars had been taught to deploy and use fire and movement. The French had not yet discarded their red trousers and kepis and their glistening cuirasses (as with the Green Howards at Ginnis, such garments were thought to strike terror into their enemies). The British khaki, on the other hand, blended into the Flanders landscape. As for the cavalry, unlike their allies and opponents, the British regiments had at last learned to fight as mounted infantry even though they still dreamed of demonstrating their prowess as the *arme blanche*. Oddly enough, one lesson the British infantry had still to rediscover was that swords (discarded just after the start of the Boer War) did nothing but make their wearers conspicuous. Several Green Howard officers died around that crossroads, sword in one hand and equally useless revolver in the other. But French officers would still go into the attack wearing white gloves and waving canes.

Early on the morning of 27 October the defences were reorganized so that 7th Division came under Sir Douglas Haig, commanding I Corps. This resulted in the 2nd Green Howards being withdrawn into reserve, but this was short-lived: that night they were back again in the forward trenches.

After two more days of intense fighting during which their colonel was among the many killed by the ubiquitous German snipers, the Battalion, now only 300 strong, was ordered to pull out in daylight from the protruding and vulnerable salient into which it had been pinned.

Only ten more men were lost in carrying out this ticklish manoeuvre, among them Corporal Tom Riordan. Son of a Green Howard RSM, he would achieve the same rank himself in the Regiment as would his own son, Jack, who went on to reach the rank of major. Made prisoner, Corporal Riordan was brutally treated by his captors, whom he saw commit many atrocities.[13] This sort of thing was not common, but it did happen. Early in the battle a party of Germans, taking refuge in a house, had refused for some time to surrender to a Green Howard patrol, even though the building was blazing around them; well treated afterwards, the Germans admitted that they had been told that the British shot all prisoners out of hand. At times this could be true: the divisional historian admitted that at one point 'there was not much taking of prisoners' during a counter-attack by a Highland regiment.[14]

The battle went on. The most critical day proved to be 31 October when Haig, followed by his mounted staff officers, at the height of the fighting trotted slowly up the Menin Road towards Gheluvelt to give confidence to the shattered survivors of units still stubbornly resisting there. Such gestures are at times fruitful. The thin line held. The next day Haig passed the message through Brigade and Division to the Battalion, 'Please congratulate the Yorkshire Regiment on their stout performance'.[15] It was typically terse and to the point. At last, on 5 November, what was left of the completely spent 7th Division was withdrawn from the line, but it was another fortnight before the Germans abandoned their attempt to break through to Ypres, that key to the defence of the Channel ports, the British Army's lifeline.

First Ypres has been likened to Inkerman, a soldiers' battle. But Inkerman lasted for hours, not weeks. Casualties in 7th Division numbered nearly 10,000, largely among the infantry; in all the British lost 58,000 men. It was the end of the old British regular army.

Someone who saw what was left of the 2nd Battalion after the battle was to write:

Never did I again see during the war men in such a sorry plight. We had lost 10 officers killed, 18 wounded and 655 other ranks killed and wounded, of

which about 250 were killed. What was as fine a Battalion as there was in the
British Army, which had started the battle about 1,000 strong, was now
reduced to one captain, three second-lieutenants and less than 300 gallant men,
but with the spirit of the old Green Howards still in their blood. The nucleus
was there on which we soon built up the Battalion again.[16]

The writer could have been Pickard, who served for the entire war with the
2nd Battalion and ended a distinguished career in the rank of lieutenant-
colonel. It could also have been Major-General Bulfin, who commanded first
a brigade and then a division in the battle and probably saw something of
his old Regiment. But he was wounded on 31 October.

Three days later the remnant of that Battalion was once again under fire,
holding a support position at the famous village of Ploegsteert – known to
all, of course, as 'Plugstreet'.

★★★

The 2nd Battalion was but the first of a succession of Green Howard
units to prove its worth on the Western Front. Next to cross the Channel
were the Territorials of the 4th and 5th Battalions. With the 4th East
Yorkshires and the 5th Durham Light Infantry they made up 150 (York &
Durham) Brigade of 50th (Northumbrian) Division, a formation that was to
become a household name in the north-east, evoking sad and proud
memories over the next thirty and more years.

The 4th Battalion, with its headquarters at Northallerton, was recruited
from the north of the Riding, the industrial complex of Teeside and the little
market towns and villages on the edge of the Cleveland Hills around
Guisborough and Stokesley. The 5th covered much of the East as well as the
North Ridings of the county – the coastal towns of Scarborough, Whitby
and Bridlington, together with such farming centres as Malton, Pickering,
Beverley and Driffield. (The Bridlington company prided itself on its descent
from the mid-Victorian Volunteer unit, the Burlington Rifles, and were still
known as such locally.)[17] With platoons drawn from a village or perhaps a
works, family pride and local spirit were to sustain these Territorials during
two world wars. Men trusted officers they might have known since boyhood,
but they were also aware that if they failed to measure up to the standards
set by their mates, word of it might reach their street or hamlet.

At the end of July 1914 the units of this York & Durham Brigade were
enjoying their annual camp in North Wales when they were ordered to pack
up and return home. Mobilizing two days later, the 4th Battalion had but
eleven absentees: one proved to be dead and three were away at sea. Enlisted
for service at home, these Territorials could be sent overseas only if they so
volunteered; when told that the War Office had decreed that if 80 per cent
of a unit volunteered, they would be allowed to embark as such, more than
90 per cent of the 4th Battalion did so. The spirit was superb.

There followed eight months of hard training, punctuated by various
alarms such as the oddly pointless shelling by German warships of their
homes in Scarborough and other coastal towns. Then, in April, they

embarked for the continent of Europe, part of the first complete Territorial Division chosen to do so.

★★★

Trained and equipped for mobile operations of the type encountered during the early months of the war, and unprepared for trench-warfare, the 2nd Battalion, together with the rest of the British Army, had endured an utterly miserable winter. By mid-November the Battalion, then located south-west of Armentières, had been brought up to strength by a large draft, more than 500 strong. Others followed at regular intervals to replace the steady drain of casualties, some from enemy fire but more from sickness, the result of a routine of three days in the line and three in reserve.

Those months were marked by torrential rain. The low-lying Flanders plain quickly became a sea of mud – that hallmark of the Western Front. As yet, however, there was little to mitigate the foul conditions. Often standing knee deep in the freezing slush of the meagre trenches, at times men had to be relieved twice daily. Boots were in short supply, as was every type of warm or waterproof clothing, although in the end goat-skin jerkins began to appear. Men went sick in their thousands, especially with 'trench feet'. Little material was available for duck-boarding or revetting the trenches and dugouts, and there was always a shortage of tools to dig them. (Reluctant as British troops ever were to use picks and shovels, the French were worse: trenches taken over from them were unfailingly abysmal). There were as yet no steel helmets. Trench mortars and grenades, the mainstay of this type of warfare, were improvised and often dangerous. There was a gross shortage of the rigidly rationed artillery ammunition, and guns heavier than the 18-pounder field artillery weapon were still few. The Germans, on the other hand, were now able to deploy the huge howitzers previously used to smash the fortifications of Liège, Namur and Antwerp. Except for the expansion and equipment of the fleet, British industry had not been geared in advance for such a war; both the will and the money to do so in peacetime had been lacking. The officers and men of the 2nd Green Howards, together with the rest of the unfortunate infantry, suffered most from this parsimony during that first wretched winter.

There was a break of sorts on Christmas Day, marked by the Germans opposite the 2nd Battalion illuminating their trenches and requesting an armistice to bury their dead. It lasted until the New Year, with both sides taking delight in walking about above ground, but, at the same time, working on their defences. Both officers and soldiers fraternised, exchanging cigarettes, cigars and other souvenirs. It never happened again. Higher authority stopped it.

★★★

When fresh units arrived in France, an attempt was usually made to teach them something about the techniques of trench-warfare by feeding them into a quiet part of the battleline. But there was to be no running-in period for the 4th and 5th Battalions.

They landed at Boulogne on 18 April 1915. Four days later British observers east of Ypres were to watch 'two curious greenish-yellow clouds on the ground at either side of Langemarck in front of the German line. These clouds spread laterally, joined up and, moving before a light wind, became a blueish-white mist such as is seen over water-meadows on a frosty night.'[18] On the British left was a newly arrived French colonial division. Quite soon, terrified Algerians were seen moving back in disorder, retching and pointing to their throats. It was the first use of poison gas, discharged from containers. Against it there was no protection, although urine-soaked cloths held over the nose and mouth helped a little. The Germans had launched this attack to divert attention from their forthcoming offensive on the Eastern Front; they were also testing their new weapon, one in which they had little faith. To their surprise its use cut out a third of the Ypres Salient, the arc around which the fighting of the previous autumn had stabilized when their earlier offensive in this area finally ground to a halt. (Because gas was a novel weapon, it was generally execrated; it also contravened the Geneva Convention.)

Holding the Salient were five divisions – two of them British, one Canadian and two French. During that first night of appalling confusion, in some extraordinary manner troops were scraped together to close the five-mile wide gap that had been torn out of the Allied defences north-east of the city. The Allies might have abandoned what remained of the Ypres Salient, but did not. At this stage of the war there were important political and sentimental reasons for holding on to that small part of Belgium still unconquered. Ypres was also a vital centre of communications.

So began the Second Battle of Ypres, that dogged struggle to recover and retain the ring of low hills overlooking the city. As the 4th and 5th Green Howards were rushed forward in motor-buses the day after the German breakthrough, they passed Belgian refugees in their hundreds fleeing from Ypres. For the first time the city had been heavily bombarded. Soon it would be a gaunt ruin.

That night 150 Brigade manned defensive positions along the line of the Yser Canal, north of the city. There, during the morning of 24 April, both battalions were first shelled, but casualties were few. Then, at about midday, they received confusingly vague instructions. They were to cross the canal, move towards the village of Potiyze, and there place themselves under the orders of any brigade commander who needed help.

There was possibly some small excuse for this. Communications had all but collapsed and the situation could hardly have been more obscure. In a succession of counter-attacks the previous day, almost every available British reserve had been destroyed. Then, before dawn on 24 April, a second German gas attack had torn a further gap in the Canadian-held sector of the front and the enemy infantry had advanced well beyond the village of St Julien, lying just three miles north-east of Ypres.

As the units of 150 Brigade moved forward towards the battle they received a medley of contradictory orders from a variety of different headquarters. Finally, however, the 4th Green Howards, together with the

4th East Yorkshires, were told first to make good the village of Fortuin and then to push the Germans back into St Julien and, if possible, beyond it. But Lieutenant-Colonel Bell, an old Volunteer now in command of the 4th Green Howards, was also warned that because troops were short 'it was inadvisable to lose many men unless some really definite advantage could be gained'.[19]

Bell was not deterred by the obscurity of his orders. Supported only in the later stages of the attack by two Canadian batteries firing over open sights, those two raw and already weary units rolled the Germans back, losing men all the time from shrapnel and machine-gun fire. They were described as 'advancing as if they were doing an attack practice in peace! They went on in such a way as if they'd done it all their lives, nothing stopped them.'[20] When the colonel and the second-in-command of the East Yorks were both killed, Bell took that unit under his command as well. Having captured Fortuin, he changed direction to reach the outskirts of St Julien. There he was told to halt. Just a week after leaving England these two Territorial units had proved their worth. It had cost the 4th Green Howards nearly 100 casualties: perhaps it is some indication of the leadership required and shown by those inexperienced officers that five of them were killed, as against ten other ranks. That same afternoon the 5th Battalion fought a similar action in which it lost sixty men in just five minutes.[21]

★★★

During the subsequent month's fighting, the 4th and 5th Green Howards moved in and out of the line, reinforcing desperate formations as best they could. For a long time they fought under the command of the Cavalry Corps, used as infantry and now so adept at trench-warfare that they helped to show these newcomers the techniques required.[22]

The culmination of Second Ypres came on Whit Monday, 24 May, when the Germans released the largest concentration of gas yet experienced. In the light wind it hung in almost stationary clouds, forty feet high. Behind and through it came the German infantry. The 4th Battalion was holding trenches astride that same Menin Road, but further back at Hooge. There, in company with the 9th Lancers and the 15th and 18th Hussars, it halted the German attack. To quote the Battalion's war diary, 'From trench to trench we fought, and thank Heaven stuck to our trenches, but at great loss'.[23] Losses that day totalled 200 officers and men. But that night the Germans gave up, having gained only a few thousand square yards of churned-up wasteland; their fighting units were as exhausted as those of their enemies and their ammunition was spent. The Second Battle of Ypres had ended. It had cost the British Army a further 60,000 casualties.

No detailed record of the casualties suffered by these Green Howard Territorials has come to light, but the 4th had lost nearly all its officers and about half its fighting strength during that month's hard fighting. The 5th were rather less unfortunate. The 50th Division as a whole had suffered more than 5,000 casualties, most of them in the infantry. Understandably, St Julien was later chosen as the site for the Divisional War Memorial.

A possibly apocryphal story relates how the Green Howards were first described as 'The Yorkshire Gurkhas' during Second Ypres. A neighbouring unit is said to have commented that the stature of these tough and wiry North Country men was in inverse ratio to their courage.[24]

★★★

While these Regular and Territorial units were fighting alongside the French during 1914 and early 1915, new formations were being raised at home. To the country's immense delight Field Marshal Lord Kitchener had joined the Liberal Government on 5 August 1914 in the civilian role of Secretary-of-state for War. His prestige, experience and stature was such that he dominated the Cabinet in the early weeks of the war; nevertheless he did accept his colleagues' advice that conscription would destroy national unity, widespread though the demand for it was. Instead, he set about raising half a million volunteers, a figure soon doubled. His call to arms for 'The First Hundred Thousand' trumpeted by 'Your King and Country Need You', appeared in the press forty-eight hours after he assumed office.

Before the war no plans had been prepared for any large wartime increase in the size of the Army. A possible framework for expansion could have been Haldane's Territorial Force, backed by its County Associations, but nothing had been done. Both ignorant and contemptuous of these amateur soldiers, Kitchener rejected all suggestions that they be used as the basis for the Army's vast expansion. After years of service abroad, he admitted that he knew neither England nor the British Army.[25] As arrogant as he was unwilling to listen to advice, he consulted the military members of his Army Council only when he pleased; incapable of delegation, he was intolerant of interference. His virtue lay in the scope of his imagination. He was one of the few who foresaw that the country must prepare for a long war. Haig was another.

Men flocked to join the long queues outside the recruiting offices. The first thing lacking was the staff to examine and enlist them. In Middlesbrough mounted police had to be summoned to help control the hundreds of men clamouring to join the Green Howards; one policeman was heard to yell, 'If you don't stop this bloody shoving not one of you will get into the Army.'[26] Nevertheless, in a masterpiece of improvisation, by 14 September the so-called 'First Hundred Thousand' volunteers had joined the Colours. A month later eighteen New Army divisions were in being – in name, at least.

The 6th (Service) Battalion of the Regiment formed part of that First Hundred Thousand. The men were 'a fine well-set up hardy lot, mostly miners, and very keen. In many cases they look upon this as the finest holiday they have ever had . . .'[27] So long as summer lasted, for the pitmen of 1914 it probably was. With a regular commanding officer of twenty years' service, the future General Harold Franklyn as its first adjutant, and a handful of other regular officers and NCOs, the 6th Battalion did not have too bad a start. The raising of the 7th (Service) Battalion for the Second Hundred Thousand presented rather more problems, not untypical of those faced by nearly all those 'Kitchener' Battalions.

Major Ronald Fife, who had retired the previous year, was recalled from the reserve to command this new 7th Battalion. His Division, the 17th, was collecting in Dorset. Towards the end of August Fife reported at Wareham. Two days later his men arrived, 1,100 of them with a single regular subaltern. The rest of the freshly commissioned officers had assimilated no more than a smattering of military knowledge in school or university OTC. There were no NCOs other than three or four elderly men who had served as such; one of them, a colour-sergeant, who had retired before the Tirah, was made RSM. What was needed was just a handful of those regular officers and NCOs who had died at First Ypres, but in the expectation of no more than a short war, none had been left behind to train their successors.

Like the 6th Battalion's, Fife's men were for the most part miners, a high proportion from County Durham, among them just a few old soldiers who had served in the Regiment; In the sparsely populated North Riding, the Green Howards were obliged to look beyond their county boundaries for many of their recruits. All were wearing their civilian clothes, there were enough tents to provide cover for about half of them, and dixies to feed one-fifth.[28] Of weapons there were none. Even at the Depot, which at one time held 5,000 men, reinforcements for 2nd Battalion were faced with a similar state of affairs. With little food and no pay to buy it in Richmond, some recruits slept blanketless on straw in the drill sheds and on the tennis courts.

The people of Wareham gave shelter to the tentless men of the 7th Battalion, as Richmond did to those at the Depot. Feeding was in relays, and Fife bought straw and eating utensils locally. Lacking any sort of nominal roll, he compiled one by drawing out the whole of his bank balance; then, helped by his subaltern (now the adjutant), he paid each man a few shillings and recorded his name. At first Fife acted as his own drill sergeant. Wooden rifles were soon available, but Fife refused to insult his men by accepting them. A chaplain of another battalion asked how a man could 'feel like a soldier with a wooden toy to carry about . . . knowing that before parade it had already poked the fire, cleared the kitchen sink, beaten the dog, or proved of domestic utility in other ways?'[29]

Working men in 1914 were brought up to poor houses, rough food and exposure to bad weather, but those who had volunteered so cheerfully were shocked by the Army's treatment of them. The pitmen who then provided so many recruits for the Green Howards made superb soldiers. Among their many qualities was their phenomenal digging; they were said to disappear underground before their task had been fully explained to them. But they did not always take kindly to military discipline. Their independence matched their sturdiness. At Christmas that year, a deputation was found squatting miner-fashion on their heels outside 6th Battalion's orderly-room. 'We've come to gie wur notice,' they announced. The reply they received is not recorded.[30]

In the end, uniforms, weapons, huts, equipment, horses, mules, vehicles and all else began to appear. On 13 July 1915 the 7th Battalion sailed for France; by the month's end it was in action. After eleven months in the shaping, an amorphous mass of men had become an infantry battalion. Only the quality of the officers and the men had made it possible. They were their country's best.

★★★

Five other New Army battalions were to be raised during 1914 and early 1915 – 8th, 9th, 10th, 12th and 13th. With one exception they all began life in the same atmosphere of neglect which was alleviated by brave attempts at improvisation by their enthusiastic young officers. Sadly, the quality of the few available regulars deteriorated. Out-of-date 'dugouts' arrived, men who had languished for years on the reserve, as well as a few elderly ex-Militia officers. Among the former was that commanding officer so censured by Harold Franklyn and now near sixty years of age; he died gallantly in France. Another battalion began life with just a major, a sergeant-major and a quartermaster-sergeant to take care of its 1,070 recruits. The resultant misery inflicted upon those fine men can well be blamed upon what has been described as the British system of 'decision by crisis'. Governments take fundamental decisions on defence matters - especially when they are likely to cost money – only when the international situation alarms the general public, and when a government feels that it has the weight of public opinion behind it.[31]

The New Army unit mentioned as having started life on sounder lines was the 12th (Middlesbrough Pals) Battalion. The widespread accounts of congestion and hardship among the newly raised units was one of several reasons why recruiting declined during that first winter. Nearly half a million men had enlisted in September. In February the total was short of 90,000. Such had been the problems of coping with the early flood of recruits that the War Office was only too glad to accept the many offers of help made by local councils, industrialists and other prominent citizens. This led to the raising of these 'Pals' battalions, in which men from a particular district or with a common background could serve alongside friends and acquaintances, a prime incentive towards recruiting.

In this way Middlesbrough's civic pride produced the Regiment's only 'Pals' Battalion. Raised in January 1915 as a pioneer unit, it went also by the name of 'The Teeside Pioneers'. Most of its recruits came from the city's skilled artisans – fitters, carpenters, blacksmiths and masons. Local bodies offered every kind of encouragement and material help, including good quality equipment made by local firms – a contrast to the sub-standard American leather obtained under War Office contracts and still worn by the 6th Battalion. In contrast to the other Service Battalions, the 12th was housed in comparative comfort at Morton Hall; there were beds for all and the YMCA furnished and ran a canteen.[32]

What was it that made these men flock in their hundreds of thousands to join the Army, not just Kitchener's battalions, but the Territorials and Regulars as well? One of them had been Second-Lieutenant A.C.T. White, who was to win both a Military Cross and a Victoria Cross in France with the 6th Green Howards. A person who counted intellect as well as gallantry among his many fine qualities, he was, in his old age, to write:

People in England fifty years ago were far more influenced by the traditional ideas of right, wrong and duty: and while few of us amongst the officers had a clear idea of the history and politics of the Balkans (which brought about the war), every private knew that Belgium had been brutally attacked, and that unless Belgium was rescued, it might be Britain's turn. It was that point, and no other, that brought the crowds to the recruiting offices: and it was that point which maintained morale until 1918, in spite of colossal casualties.[33]

There were, it must be said, other factors. An escape from boredom, added to a sense of adventure, was undoubtedly one. The heady atmosphere engendered by the recruiting campaign influenced many, especially those whose friends were joining; impulse governed others. In the early years, before industry and agriculture were put on a war footing, an escape from unemployment or short-time working could push men towards the recruiting office. Overlying everything else, however, was a simple patriotism common to all classes, that strong sense of duty and obligation emphasized by Archie White. Others overseas were moved by it as well. From every part of the British Empire, Canadians, Australians, South Africans (Boer and British), New Zealanders and Rhodesians clamoured in their thousands to come to the aid of what they called 'The Old Country'.

***

Before the war the French Army had sought the key to victory against the time when it could take its revenge for its previous humiliation by Germany. In Napoleon's battle-winning philosophy of *l'attaque, l'attaque, toujours l'attaque* – the headlong and all-out offensive regardless of cost – its generals thought they had found it. Its virtues, indoctrinated at every level, were to cost the French nearly a million casualties during 1914 and another million and a half in 1915. The lessons that might have beeen learned from the American Civil War and the Russo-Japanese War of 1905 were ignored: magnificent *élan* could make little headway against barbed-wire, the machine-gun and massed artillery.

To help create diversions on the battle-front, the French agitated for their British allies to attack in the same manner, small though their numbers still were and meagre their resources. The 2nd Green Howards whom we left at 'Plugstreet' Wood when 1914 was drawing to its close, was among those who suffered in this way. On 10 March 1915, after having endured that miserable winter, it took part in the first British offensive on the Western Front, an attack on the village of Neuve Chapelle. The preparations by Haig's First Army staff were well conceived and meticulous; surprise was gained. The first assault was a striking success, but thereafter the attack bogged down. Communications collapsed: vital field-telephone lines were time and again destroyed, runners quickly hit. Reserves, among them the Green Howards, were delayed among a maze of smashed trenches and water-filled drainage ditches. A confused struggle ensued over the next three days during which Corporal William Anderson, leading a small party of bombers, won the Regiment's first Victoria Cross of the war – posthumously as so many were. On the third day of the battle the Battalion helped halt a major counter-attack that left hundreds of German corpses lying in front of its trenches. But it all ended with the capture of one smashed village at a cost of 13,000 British casualties, more than 300 of them Green Howards, one of them a future Colonel of the Regiment, Lieutenant A.E. Robinson.

Paradoxically the battle had done much for the British Army's reputation – not just among its French allies but with the Germans as well. In future the latter never neglected that part of the front held by the British, while the French realized that they might even learn something from British methods. It was also encouraging that the misery of months of routine trench-warfare had little dampened either the daring or the initiative of units such as the 2nd Green Howards, a battalion that had been bought back to strength again and again by a succession of virtually untrained drafts both of officers and other ranks. (One man remembered that, when he arrived in France in November, he had *handled* and fired a rifle only once; he was wounded within two days.)[34] Although the Battalion was now regular in little but name, many of its standards survived and continued so to do throughout a further three and a half years of war.

Such profitless operations were the pattern of life for the British Army during that spring and summer. At Festubert in May initial success was followed by stalemate. It cost the Battalion another two hundred killed and wounded, including their fine commanding officer, Lieutenant-Colonel W.L. Alexander, who had been wounded at Ypres and had rejoined. At Givenchy in June losses were even worse, more than 400 in all, the heaviest in the Division. An understated account survives of how one 'A' Company soldier spent a warm and sunny Sunday morning:

> We laid in our trenches listening to our Artillery shelling the Germans, but for every shell we sent over we seemed to get about four in return. Small wooden ladders had been placed behind the parapet and on the whistle sounding we all scrambled up these ladders and 'over the top'. We were met by lots of machine-gun fire and men were falling right, left and centre. The barbed wire had been partially destroyed but it still caused considerable trouble. I was one of only ten men who got to the German first line trench. My Platoon Sergeant (Sgt Whitlock) was one of the ten and he quickly got us organized. We naturally expected reinforcements but did not get any. All the Company officers had been killed except Mr Belcher, and he was badly wounded.'[35]

Whitlock, who pulled the survivors back when it was dark, was awarded the DCM. The writer also was seriously wounded. In all, seventy of the 360 members of two assault companies got back. Givenchy was a complete failure.

The Battle of Loos was a repetition of those previous attacks, but on a larger scale. In 25 September yet another major Allied offensive was launched, its purpose to relieve pressure upon the Russians, seemingly near to defeat, and to take advantage of the German preoccupation with that Eastern Front. Forty-three French and a dozen British divisions were committed, the latter among the wrecked villages and slag-heaps of the mining area around Loos. Both Kitchener and Sir John French, the British Commander-in-Chief, supported the French in their insistence on mounting yet another such offensive. Haig, whose First Army was due to attack with insufficient guns and less ammunition, was among those who strongly opposed it.

Loos was marked by the first appearance on the Western Front of some of Kitchener's New Army divisions, the 21st and 24th. The former included the 10th Green Howards. Inexplicably, these two raw formations were included in the GHQ reserve, the task of which would by its nature inevitably prove complicated. When this reserve corps was released to help First Army exploit its initial success, the chaotic experiences of the 10th Battalion matched those of many other New Army units.

The initial orders, given to the battalion commanders off a 1:100,000 map, were vague enough – to take Hill 70, east of the captured village of Loos, if the Germans still held it. By way of congested roads the Green Howards threaded their way by an uncertain route towards the din of the battle they could hear raging ahead; they still wore their cumbrous packs and behind travelled their battalion transport. At length they came in full view of the German artillery observers. In no time the remains of the unit's vehicles and animals completely blocked the road. Without any guides and never having seen the country before, they were not alone in failing to locate Hill 70 and so they went badly astray. Ignoring the warnings of one of the forward units, they moved straight on across its trenches to come under heavy machine-gun fire. It was an inopportune start, its cause bad staff work and inexperienced regimental officers. Among the endless shortages with which formations tried to cope was that of trained staff officers.

After these and other vicissitudes the 10th Battalion was one of a number of units that tried the following morning to capture the German redoubt built on Hill 70. Under the hail of shells and bullets the attacking lines, almost devoid of artillery support, seemed to wither away. In a final attempt to encourage their men forward, in the words of the Official History, 'First Colonel Hadow, got up and rushed forward shouting "Charge!", and was killed; then Major W.H. Dent rose and did the same, and was killed, and the next two senior officers met the same fate in the same manner.' High-spirited gallantry was almost the sole quality of those almost untrained units. Only one officer of all the 4,000 officers and men of their brigade was said to have had any previous experience of war.[36]

All along the line it was much the same story. At last, towards the end of the day, men began to move towards the rear, at first in an orderly manner and then in a rabble. Too much had been asked of these fine troops, the pick of the country's volunteers. In a jumble of units some men stayed put – hungry, wet and thirsty; amid the racket and confusion, platoon commanders hung on without orders. Battalion casualties in the end numbered more than 300.

Fighting just to the north of this New Army unit was the 2nd Battalion who, at the start of the battle, had suffered from their own chlorine gas, first used that day by the British on a large scale, blowing back in their faces. After a week's heavy fighting, their losses matched those of their 10th Battalion. In all, British casualties at Loos numbered more than 50,000. Nothing had been gained except valuable experience, to be made sound use of in the future. It was an expensive way to buy it.

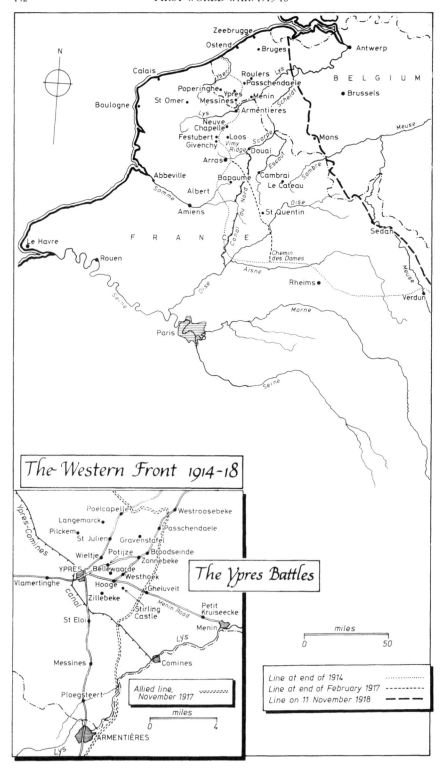

The Western Front 1914-18

The Ypres Battles

Allied line,
November 1917

miles
0   4

miles
0   50

Line at end of 1914 ....
Line at end of February 1917 ------------
Line on 11 November 1918 ━━━━━━

7. First World War: 1915-18

Although he was writing seventeen years after it all occurred, the Official Historian still managed to catch the atmosphere at Suvla Bay on the night of 6 August 1915:

> Not a sound came from the land. Fleecy clouds hid the stars. A slight breeze ruffled the quiet sea. At this beach the prevailing conditions were ideal. The beach was undefended. The lighters grounded within a few yards of the water's edge; their ramps were lowered on to dry sand; and from all seven lighters the troops poured ashore two abreast exactly in accordance with the plan.[1]

Two of these lighters, the so-called 'Beetles', carried the 6th Battalion of the Yorkshire Regiment. This secret night landing on the west side of the Gallipoli Peninsula was notable as being the first occasion upon which a New Army unit had been committed to a large-scale action. Loos had not been the first major battle fought by a 'Kitchener' Battalion of the Regiment.

***

Turkey, snatching at the opportunity of revenging herself against Russia, her almost hereditary enemy, had thrown in her lot with the Central Powers in the late October of 1914. Her entry into the war caused consternation in Britain, the people of whose Empire included many millions of Muslims, among them some of the finest soldiers of her Indian Army. Further cause for concern was the threat posed to the safety of the Suez Canal by the Turks, rulers of the greater part of the Arabian Peninsula. When Russia appealed for help against Turkey, an expedition to force the narrow straits of the Dardenelles appeared to be a sound undertaking. Leading to the Sea of Marmora and Istanbul, then known as Constantinople, their capture could lead to the opening of a sea route to southern Russia and might also persuade a wavering Italy to enter the war on the side of the Allies. The Turks might even seek peace.

So it was that on 25 April 1915 an Anglo-French force that included two recently raised divisions of Australians and New Zealanders carried out a successful opposed landing on the Gallipoli Peninsula, on the north side of the Straits. But the country was rugged, water scarce, the heat overpowering and the Turks, advised and sometimes commanded by German officers, excelled at defensive fighting. Their reinforcement and supply bases were also nearby. A summer of dour fighting lay ahead, one in which the Allies would barely expand their initially small bridgeheads.

This Gallipoli campaign produced the first major clash of the war between the 'Westerners' and the 'Easterners', two opposing schools of strategic thought. The Westerners were sure that the war could be won only by destroying the armies of Germany on the Western Front, their principal and most redoubtable enemies. The 'Easterners', horrified by the hundreds of thousands of deaths that had already occurred in attempts to break through the flankless fortifications stretching from Switzerland to the Channel, sought a cheaper alternative, one that would make full use of the mobility offered by the combination of superior British and French sea power. The 19th of Foot had experienced this 'strategy of evasion' (as the prominent military writer, Major-General J.F.C. Fuller, christened it),[2] at Cadiz, at Belle Isle and often in the West Indies.

★★★

During June and July substantial reinforcements had been received by the Commander-in-Chief, General Sir Ian Hamilton. (It was something of a link with the past, that he had made his reputation on the North-West Frontier while acting as ADC to that Crimean veteran of the 19th Foot, 'Redan' Massy, transformed into a cavalry brigadier). Hamilton's plan was to use two of these fresh divisions, 10th and 11th, to land at Suvla Bay in yet another attempt to break the Gallipoli stalemate. Simultaneous attacks would also be made from the existing bridgeheads further south at Anzac Cove and Helles. From this Suvla landing the hills at the narrow northern neck of the peninsula could then be seized, so severing from its base the main Turkish army. The likely opposition at Suvla was known to be small.

Lala Baba, a 200-foot-high hill, commanded the southernmost arm of Suvla Bay and overlooked a dry salt lake which extended behind it towards the higher range of hills beyond. The role of the 6th Green Howards was to take Lala Baba and a smaller hill on the far side of the salt lake, a formidable task for such a half-trained unit. The men were already weary when they landed, having been on their feet for seventeen hours. They had also just received a cholera 'jab' whose effects in those days were similar to a kick from a mule. The previous day a hot march to the embarkation point had followed a tiring morning's training. Through an ill-judged attempt to maintain security, commanding officers had been kept in the dark as to what was afoot. Not until the early afternoon did company commanders learn what was happening.

Scrambling ashore from their 'Beetles', the primitive forerunners of the landing craft that were to be used by this same battalion in Normandy twenty-nine years later, the two leading companies formed up and quietly moved off towards their objectives even before the rest of the unit was ashore. Their orders were to use only their bayonets until daybreak. Trudging up the lower slopes, they could just see the rounded hill-top. Suddenly a red flare exploded and a storm of bullets lashed them. Confusion followed, as in most night attacks, but these New Army troops, raw as they were, kept moving up the hill. Some Turks lay low as the Green Howards passed over their trenches, only to leap to their feet and attack with bayonet and bullet from behind, but by midnight the Green Howards had cleared Lala Baba. Helped by a West Yorkshire battalion that landed behind them, they had taken the sole enemy position commanding the landing beaches. In so doing every officer but three had been either killed or wounded. One-third of the men had been hit. The colonel, his second-in-command and two of his company commanders were either dead or dying.

An extraordinary inertia then overtook the assaulting divisions. Two full days were wasted before the corps commander tackled the main hills in the distance, but by then it was far too late; the Turks had brought up strong reinforcements. The concept of the operation had been sound. The operational plan had been adequately put together by the inexperienced staffs. The landing was near faultless. But in a paralysis of the command system, the subsequent execution failed utterly.

Hamilton, directing operations personally from the nearby island of Imbros, was a sensitive and talented Scot, intelligent and forward-looking. He was able to cope with French and Australian allies and with sailors and politicians. But he was sixty-two years old and failed to see that age could be a fault in lesser men. His subordinates in those New Army divisions that dawdled at Suvla Bay were elderly and inexperienced 'dugouts', often unfit into the bargain.

That summer of 1915 was the nadir of First World War British generalship. In subsequent years little mercy was to be shown to those who failed. Haig was to sack a hundred generals after he took over command in France from Sir John French at the end of 1915. But it had all happened before and was to happen again, that reluctance to replace elderly and inflexible generals at the outbreak of a war. In 1914 such men had even been brought back from retirement to be given command of field formations.

***

Within the next two weeks, three more divisions and a force of dismounted cavalry were fed into the Suvla bridgehead, but this mass of extra troops, short as they were of guns, could do little but smash their heads against the now strong Turkish defences in a series of costly and abortive attacks. Reinforcements for the 6th Green Howards were also landed, among them Captain Archie White, who for the time being took over command of his depleted Battalion. He had been among the sick left behind at Imbros. His brother had died in that bloody night attack.

Thirst, flies and a diet of warm bully and biscuits were the 6th Battalion's main enemies for the next fortnight, spent mostly in the line, although Turkish snipers took a steady toll. They then took part in the final attempt to break out of Suvla. Acting under orders from their brigade headquarters which succeeded in being both vague and complicated, in mid-afternoon on 21 August the Green Howards once again went over the top. The preliminary bombardment from heavy naval guns firing armour-piercing shells had sounded encouraging, but had done little or no damage to the well-dug and concealed Turkish trenches. The Battalion was to have moved on a compass bearing across the flat and featureless plain, but the officer using

the single available instrument was soon mortally hit. Of the other eight
officers, five died and three were wounded. Under a company sergeant-
major, about a hundred men kept going to reach the Turkish first line. No
reinforcements or orders ever reached them. The next day the survivors fell
back to their starting-point.

As White afterwards wrote, by comparison with these Turks the New
Army soldiers were still amateurs.[3] Even more amateurish was the staff
which was utterly incapable of controlling large bodies of troops once the
inevitable fog of war enveloped the battlefield.

For a time the Quartermaster commanded the 6th Battalion which now
had only 258 other ranks. Then, as reinforcements arrived, it settled down
to the nastiness of trench life, a wearying routine punctuated by sudden death
and injury. But the futility of continuing the campaign had become clear,
and Gallipoli was evacuated. The paradox was that the withdrawal was
superbly executed. A total of 134,000 men were shipped out without one
being lost, the 6th Battalion leaving five days before the Christmas of 1915.
For many years the Staff College studied that withdrawal, a model operation.

Half a million men had tried to capture Gallipoli; half of them became
casualties. The losses matched those of the Turks, a mark of the intensity
of the fighting. Time and again the Allies had been close to success, but the
ineptitude of the generalship and staff work had equalled that of the Crimea.
But whether victory at the Dardanelles would have brought about the capture
of Constantinople, the opening of the Bosphorus to supplies for Russia, or
even to the collapse of Turkey will always remain unanswered.

Another casualty of Gallipoli had been the campaign's main proponent,
Winston Churchill, the First Lord of the Admiralty. He left Whitehall to
fight as an infantry officer in France, his reputation for strategic perception
seemingly irreparably ruined. The prestige of Kitchener, another enthusiast
for the undertaking, suffered a similar blow, one that could have led to his
removal if he had not been drowned when HMS *Hampshire* was sunk while
taking him to Russia in the following year.

***

During that summer of 1915 three more New Army units of the Green
Howards had arrived in France, the 7th, 8th and 9th, bringing to seven the
total of their battalions facing the steady attrition of the second autumn and
winter on the Western Front. There was now time to introduce raw
battalions rather less harshly to trench life by first giving them experience in
less active sectors of the line. However, artillery fire and snipers were soon
taking a steady toll, and from time to time units were involved in minor
attacks, minor that is to everyone but those taking part. Fife could describe
a day as 'fairly quiet' in which the 7th Battalion lost only eleven men.[4]

The effect of the winter rain on the pulverized ground, especially in the
low-lying areas of Flanders, was unspeakable. Proper dugouts were still rare.
For days on end men stood, ate and fought knee deep in mud and slush; to
this misery was added the disgust of being lice-ridden. Only at night, and
with infinite labour, could luke-warm food be dragged up to the forward

troops. Everywhere lay rotting corpses of men and animals, either unburied or disinterred by the guns. Over everything was that same loathsome stench, a compound of death, high-explosives and excreta. And everyone was always utterly tired; after a week in the trenches, the move back to reserve brought to already exhausted men yet more carrying parties, more digging and wiring, and more fatigues. Nor, when the troops came out of the line, could they enjoy any proper amenities: the NAAFI had yet to see the light of day and it was not until 1916 that the first concert parties came on the scene. But officers did what they could to keep their men happy. The word 'welfare' was, however, as yet uncoined.

Archie White remembered how living conditions in the Dardanelles had been no different from those in Wellington's Peninsular Army; in 1915 the Western Front was little better. In later life he was not alone in wondering how morale could have stayed so high in such conditions and whether later generations would have stuck it in the same way. Most of the men came from farms, factories or mines and were accustomed to a measure of hardship; the many better-educated members of the middle classes who found themselves living under the same conditions adapted.[5] Much the same applied to the armies of Germany, France and Russia, all of which contained an even higher proportion of tough, sturdy country peasants. In White's words, 'The more civilization brings comfort, the harder it is to go soldiering.'[6]

It did not take long for those Territorial and New Army Green Howard units to shed their rawness in such conditions, a process that was helped by the arrival of a sprinkling of regular officers and NCOs of the 2nd Battalion, men who had recovered from wounds inflicted at First Ypres or Neuve Chapelle; with them were a number of senior 1st Battalion NCOs brought home from India. But the shortage of competent professionals at all levels was still serious. Perhaps more than anything it affected tactics. Platoons knew how to defend their stretches of trench, and mount a quick counter-attack or patrol, but a laborious rigidity crept in when anything on a larger scale had to be attempted. Because senior commanders and their staffs knew how little expertise existed at lower levels, they piled detail upon detail in their orders, leaving little to the initiative of their juniors.[7]

***

In December 1915 the Allies had agreed to mount simultaneous offensives the following summer in their different theatres. But the Germans anticipated the French, the Austro-Hungarians the Italians. The Russians alone got their blow in first, the so-called 'Brusilov Offensive' which inflicted a crushing defeat on the Austrians; so costly was it in casualties to both sides that it was to be a major factor in the break-up of both their empires.

On the Western Front the French and British were to have attacked together astride the River Somme, but the Germans struck first, mounting at Verdun in February 1916 their first major offensive on the Western Front since the opening stages of the war. For ten long months the German and French armies, on opposite sides of the River Meuse, hammered each other to pulp. In front of Verdun 700,000 men were to be mutilated or slain, most of them by concentrated artillery fire. It was attrition at its most horrendous.

By the summer of 1916 fifty-eight British infantry divisions, numbering more than a million and a half men, were in action on the Western Front, the Army's expansion an organizational triumph, but one little appreciated by their Allies. However, its lack of training and consequent tactical rigidity were to cost it dear. On 1 July, across the rolling chalk downlands of the Somme, some 100,000 infantrymen from eleven British divisions climbed out of their trenches to assault the German first line. By nightfall 57,470 of them were dead or wounded. The seven-day preliminary bombardment of more than a million and a half shells had not been enough to smash the German defences dug deep in the chalk, or smother his artillery. No more than slight dents had been made in the enemy line.

The fate of the 7th Green Howards in the 17th Division was typical of that of many fine New Army battalions. Its attack on Fricourt village was planned to start after other battalions of its brigade had breached the enemy line. This initial attack failed, a battalion of the West Yorkshires being almost annihilated. As a consequence the brigade commander attempted to cancel the Green Howards' attack but to no avail. Already a tragic error had occurred. That morning 'A' Company of the 7th Battalion had attacked prematurely and been destroyed by a single enemy machine-gun within twenty yards of leaving its assault trenches.

At 1430 hours the rest of the Battalion met a similar fate. In front of them no more than four small gaps had been cut in the wire; the deep dugouts in which the Germans had sheltered during the shelling were quite

Above: *In Delville Wood, 1916.*

Below: *The attack on the Somme, 1 July 1916.*

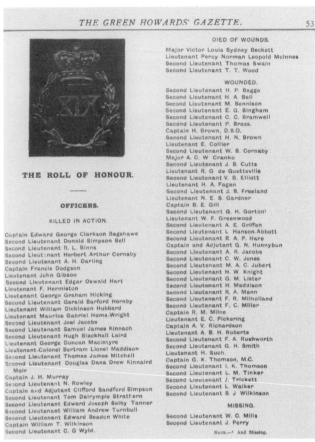

Left: The officer toll of the first few days of the Battle of the Somme. A page from the Green Howards' Gazette *which, by the summer of 1916, had been reduced by the censor to little more than a list of casualties and awards.*

untouched. Even as the men began to clamber over their parapets they could see the German trenches bristling with bayonets and lined with steel helmets; some of the enemy were standing on their parapets as if on a shooting range. Within just three minutes thirteen Green Howard officers and more than 300 other ranks had been mowed down. Only a minute handful of these superb amateur soldiers reached the enemy wire. A single man got to within a few yards of the enemy trenches, where his body was eventually found, unthrown bomb in hand. Lieutenant-Colonel Fife, still in command, had survived, only because losses among senior officers had been so severe that battalion commanders had been strictly ordered not to move forward from their headquarters until the enemy first line had been captured. Three days later he underwent the horror of walking across the captured ground through the corpses of his Battalion lying in their neat lines in front of the fortified village.[8]

In the preliminary fighting that day just north of Fricourt, of a single company of 21st Division's 10th Green Howards only twenty-eight men survived; its commander, Major S.W. Loudoun-Shand, was awarded the Victoria Cross, but posthumously. Farther south 30th Division, to which the 2nd Battalion had been transferred a few months earlier, attacked Montauban Ridge but with greater success, capturing its objective, taking large numbers of prisoners and cutting a salient out of the German position. But the problem of exploiting a breakthrough in the enemy line was still insoluble. The first handful of ponderous and unreliable tanks had only just appeared; horsed cavalry was valueless; artillery, the queen of the 1914-18 battlefield, could quickly seal off any breaches in the front as soon as they were made.

In its initial assault on 1 July, the 2nd Battalion lost 200 men. By now, however, a battalion went into battle leaving with its first-line transport anything up to one-third of its strength to make good the inevitable losses. This meant that this regular unit had enough men to attack once again on 7 July. That day it accomplished nothing, its casualties even higher than they had been six days before. Reinforced by a dozen subalterns and 300 other ranks, drawn from a variety of different regiments, two weeks later it was pushed in again, to lose yet another 250 men.

And so the Battle of the Somme dragged on until November in a grim agony of attrition with division after division committed to the fighting, and many units attacking time and time. Every Green Howard unit on the Western Front was sooner or later drawn into his hellish carnage: the now seasoned Territorial Battalions of 50th Division, the 4th and 5th; the 12th and 13th, part of 40th Division fresh from home; and the 6th Battalion, brought back from the Mediterranean with 11th Division. In two bitter weeks during September the 6th was again torn to pieces, taking part in a succession of assaults and bitter defensive actions against counter-attacks, in one of which Captain Archie White was to win yet another Victoria Cross for the Regiment. His commanding officer, Lieutenant-Colonel C.G. Forsyth, an immensely popular, highly efficient and independently minded regular, who had remade the unit after Gallipoli, was killed; at the age of twenty-eight he was already earmarked for command of a brigade.[9] One of six brothers, five were to die before the war was out.

Two more Victoria Crosses were to be won by Green Howards on the Somme, both recipients serving in the famous 23rd (Northern) Division. That to Private William Short of the 8th Green Howards was tragically posthumous as it also was to Second-Lieutenant Bell of the 9th Battalion. But as Lieutenant-Colonel Fife remarked, most of the sixteen individuals he recommended on one occasion for gallantry awards would, in the old days, have qualified for a Victoria Cross.[10]

In four and a half months the Battle of the Somme cost the two Allies some 600,000 casualties, more than two-thirds of them British. French losses at Verdun probably numbered about 350,000. Haig has in retrospect been harshly criticized for continuing the offensive in the face of such shocking losses, but German, French and British generals alike were convinced that the war would be won only by wearing the enemy down. Revolting though such a concept of attrition might be, given the difficulty of exploiting any breakthrough it was the essential preliminary to victory. And all the time Joffre never slackened his demands that Haig should continue attacking so as to take pressure off the French struggling at Verdun.

Exact figures for enemy losses on the Somme are not known, but the outcome of the battle was the German withdrawal to the Hindenburg Line early in 1917. Afterwards General Ludendorff, later the directing brain behind the German Army, was to confess that it had been fought to a standstill and was never the same again. As John Terraine has written, 'What the Germans had intended to do to the French at Verdun, the British did to them on the Somme.'[11] To say that the cream of Britain's youth died on the Somme is hackneyed, but an immutable truth never to be forgotten. It was the sacrifice of men such as those Middlesbrough and Scarborough volunteers in the New Army Battalions of the Green Howards that made possible the eventual Allied victory, although few of the survivors saw it as such, either then or later. They had also learned much. After the Somme their units and formations were to become professional, copying the infiltration tactics used by both the Germans and French at Verdun and developing high standards of co-operation between the infantry and its supporting artillery.

That winter, as the fighting died away, the weather was to prove even more vile than in 1915/16, the hardest for twenty years. Archie White recalled that some of his most wretched memories were of those months. The makeshift trenches were a compound of chalky slush, snow, water and battle debris.[12] Until everything froze solid at Christmas, the 7th Battalion saw:

A horrible scene of desolation. The whole landscape was one sea of mud and slime. Dead men and fragments of men lay everywhere . . . A lowering winter sky and driving rain completed the picture.[13]

A general was to complain to Fife that some his officers were not looking very cheerful; he hoped that they would try and look brighter. Fife's reaction was to contemplate detailing an officer in each company as a professional buffoon with orders to put on an act as soon as that general hove in sight.[14]

Men took their chance in the open rather than risk drowning in the mud of the communication trenches. Sickness mounted, but inoculations had now checked the old curses of cholera and typhoid; it was the less serious diseases such as lice-borne trench-fever and trench-feet that steadily thinned the ranks. A fatalistic pertinacity had replaced the earlier crusading heroics. Herbert Read, poet and very gallant Green Howard, in his lines from 'The Happy Warrior', caught the new savage mood:

His wild heart beats with painful sobs,
His strained hands clench an ice-cold rifle,
His aching jaws grip a hot parched tongue,
And his wide eyes search unconsciously

He cannot shriek

Bloody saliva
Dribbles down his shapeless jacket.

I saw him stab
And stab again
A well killed Boche.

This is the happy warrior,
This is he.[15]

Conscripted soldiers were now arriving in France. As early as the end of 1914, the flood of volunteers had slackened, although places like Middlesbrough could, as late as the summer of the following year, raise first its 12th (Pals) Battalion for the Regiment, followed quickly by yet another, the 14th, a reserve for the former. Abhorrent though conscription was to the British, slowly through 1915 the principle of compulsion had gained acceptance, and early in 1916, for the first time in the country's history, a formal system of national conscription for service overseas was introduced, limited to single men between the ages of eighteen and forty-one, but soon widened. The Irish, however, were exempt. Very soon units in France and Belgium were manned largely by these conscripts and by wounded men returning to the carnage for the second or third time.

***

The year 1917 opened with Nivelle replacing Joffre in supreme command of the French armies. The former, a plausible individual, offered an alternative to attrition: outright victory would be assured within a space of forty-eight hours by a single massive offensive on the Aisne, this to be supported by a subsidiary attack by the British at Arras, on the opposite and northern flank of the devastated Somme battleground. The result was disaster. The German 25-mile withdrawal to their newly built Hindenburg Line upset the French planning and yet another over-ambitious offensive collapsed against German concrete, wire and machine-guns. Many of Nivelle's until then indomitable *poilus* mutinied. The main burden of the war was afterwards to be carried by the British Army.

The Battle of Arras was just that little more successful than the Somme. Unlike the latter, ammunition was plentiful for the 3,000 guns massed in support, now manned by highly efficient artillerymen. The planning had in every way improved, as had the tactics of the infantry; the British Army was at last becoming an effective fighting machine.

Four battalions of the Regiment were to fight at Arras. First into action was the 2nd Battalion, attacking on 9 April, the opening day of the offensive when the Canadians captured the dominating Vimy Ridge and ripped a gap,

Below: *The successful local attack in frozen snow by the 7th Battalion on 8 February 1917 at Sailly-Saillisel; but they lost 60 per cent of an already reduced strength of 330. Photograph of an unknown painting.*

Above: An unusual photograph of the 5th Battalion in March 1917.

in places four miles deep and double that in width, through the German defences. The 2nd Battalion suffered the now standard heavy casualties, but two weeks later was again capable of being launched against the German defences. Fighting alongside the 4th and 5th Battalions of the Regiment in 50th Division, at one point it relieved the 5th in its stretch of trenches.

Yet another tragic day for the 4th Battalion was 23 April when the last surviving officer, Captain David Philip Hirsch, a member of a well-known West Riding family, was first wounded and then killed commanding the little that was left of his unit in the forward trenches. His was to be yet another posthumous Victoria Cross. Another such decoration, the first his division had received, was won by Private Tom Dresser of the 7th Battalion on 12 May. By then Arras had degenerated into the habitual slogging match. After six days' fighting during that second week of May, of all the 7th Battalion rifle companies' officers, only a single subaltern remained. By then the Battalion had also lost Lieutenant-Colonel Fife, its magnificent commanding officer, at long last severely wounded. Of the thirteen infantry battalion commanders who had landed in France with the division nineteen months earlier, he was the last to be either killed or wounded.

★★★

With the French Army mutinous, Russia in revolution and the successful German U-boat campaign threatening Britain's lifelines, the outlook for the Allies in the summer of 1917 was grim indeed. Hope for the future lay in America's declaration of war in April. But she was in no way ready. Somehow defeat had to be staved off until the full strength of the United States could be brought to bear, her minute peace-time army expanded and equipped. For the rest of 1917 the troops of the British Empire would struggle on virtually alone.

So it was that the Third Battle of Ypres, generally known as Passchendaele, was mounted, its main objects first to prevent the Germans from attacking the tottering French armies and secondly to remove the menace of the U-boat bases at Ostend and Zeebrugge. Furthermore there was the hope that the capture of the railway centre at Roulers, only twelve miles from Ypres, could well lead to the total collapse of the German defence of Western Flanders. Haig would break through if he could. Otherwise continued attrition would wear down the German Army.

A preliminary to the main offensive was the capture by General Sir Herbert Plumer's Second Army of the Messines Ridge, the feature which dominated Ypres and the British positions south-east of that city. Before dawn on 7 June 1917 nineteen mines containing a million pounds of high-explosives erupted under the Germans defences. Covered by the fire from more than 2,000 guns, which in a week fired more than three million shells, 100 British, Australian and New Zealand battalions began to assault the Ridge. They were well trained, well rehearsed and well supported. New scientific tools had become available: artillery flash-spotting and sound-ranging; aerial photography (air superiority over the battlefield had been won beforehand); even primitive and unreliable wireless communications for the upper echelons of command.

The most northerly of the assaulting formations was the 23rd Division and among its battalions was the 8th Green Howards. The moment the mines exploded, its men rose to their feet and began to climb the slope, their task to take the ill-reputed Hill 60, the scene of savage fighting for two long years past. Encouraged by the immensity of the explosions, within ten minutes they had occupied its now cratered summit; ten minutes later Battalion Headquarters was established on Hill 60's eastern slope and the reserve company was pressing on up towards the Battalion's third and final objective. By the day's end losses were heavy, more than 250 officers and men, as were those of its sister unit, the 9th Battalion, part of the brigade reserve, held up by surviving German machine-gunners around the final objective. It was the price to be paid for the most conclusive British success of the war so far. By mid-afternoon the vital Messines Ridge had been captured. Fighting continued for six more days during which a succession of German counter-attacks were crushed. In this the 6th Battalion was involved but details of what happened are sparse.

In the end Messines cost the British Empire 25,000 men, by First World War standards not too high a price. Germans casualties were much the same. The rather surprised delight with which the British press greeted Messines was similar to its coverage of the equally unexpected victory at Alamein twenty-five years later.

***

After the essential preliminary of Messines, Third Ypres proper opened on 31 July with an attack by sixteen divisions of General Sir Hubert Gough's Fifth Army, supported by twelve of Plumer's on his right; six French divisions were deployed on Gough's left and a further four British divisions along the coast.

In steadily worsening weather, the fighting was to continue until 12 November. Much has been written about a battle whose horrors were encapsulated in Siegfried Sassoon's famous lines:

> I died in hell –
> (They called it Passchendaele)

As the guns smashed the delicate Flanders drainage system, the countryside relapsed into a near impassable porridge of black viscous mud which drowned those who slipped or were blown off the duck-boarded tracks. Only with painful slowness could the reserves and the artillery drag themselves forward.

The first day opened quite well with Gough's main attack thrusting two miles into the German defences. On the right, up on the Gheluvelt Plateau, a patchwork of shattered woodlands, the 2nd Green Howards was in the leading assault. Despite German resistance from their 4-feet-thick reinforced concrete pillboxes, the Battalion advanced 1,000 yards to take its objective. But the reserve units failed to make any further progress and suffered heavy losses in trying to do so. By midday it was clear that the corps attack had failed.

In the words of their commanding officer, 'The 2nd Battalion may have been lucky to have been the first to go over the top . . .'[16] Not only did the Green Howards gain their objective, but they stayed on it for four terrible days. By the time they were relieved the battle had cost them what seemed to be the average for such work – something like half the battle strength of each rifle company. For them, however, it was the end of the battle, their division being now moved to a relatively quiet area of the front.

This failure to gain the high ground around Gheluvelt was to have disastrous results. It had begun to rain on the afternoon of 31 July and, with short breaks, the downpour was to continue for a full month. Next into action, on 14 August, was the 6th Battalion, attacking in the morass that had been the left bank of the Steenbeek, the most impassable area of the Ypres Salient. This attack was but one of Gough's many fruitless attempts to press forward under observation from the Gheluvelt area. In and out of the line, becoming ever more weary, went the 6th Battalion, losing ever more men.

Dissatisfied with Gough's handling of the battle, Haig handed over its control to Plumer on 25 August. Halting operations for three weeks so as to make the meticulous preparations which were his hallmark, Plumer planned a four-step operation, first to capture the Gheluvelt plateau and then the main ridge on which lay Passchendaele village.

In the first of these steps, to be known as the Battle of the Menin Bridge Road, both the 8th and 9th Battalions were engaged. The operation was a complete success, as was the part played by the Regiment, fighting alongside the Australians. The victory owed much to fresh infantry tactics, practised before the battle. Lines of widely deployed skirmishers pinpointed defences undiscovered by the supporting artillery. These were then dealt with by small groups of Lewis-gunners, rifle-grenade men and bombers; meanwhile other small groups infiltrated between the strongpoints, and mopping-up parties dealt with anything that had been missed. But the cost as ever was heavy, especially among the Green Howards. The 9th Battalion put in its final assault with a single officer in each rifle company still on his feet. German losses were probably even worse. For the Chief of Staff of the German Army Group opposite, 'The Hell of Verdun was surpassed.'[17]

Plumer's second hammer-blow by fresh divisions on 26 September was equally successful, as was his third on 4 October, the Battle of Broodseinde, for which 21st Division's 10th Battalion had been brought up from Arras. This battalion's understated record of its appalling experiences at Broodseinde, in which the men fought up to their knees in mud and water, and 'the task of the stretcher-bearers was rendered extremely difficult'[18] demonstrates how what seems to be muddled chaos on the ground can form part of a clear-cut victory – a victory that cost the 10th Green Howards 334 casualties.

Although the weather became even more vile and the ground even more impassable, Haig pressed on with the offensive against the advice of both Gough and Plumer. The fourth step, undertaken on 9 October, was to be yet another ordeal for 6th Green Howards, one that cost the Battalion dear, achieved little or nothing, but was marked by the award of its second Victoria Cross, posthumously to Corporal William Clamp.

This attack did not signal the end of the fighting for either the 6th or the other units of the Regiment still in the Salient; these were now joined by the 7th Battalion, also brought up from the Arras area. The latter spent a mere sixteen days in the line there, mostly in support, but described as 'one of the most horrible times spent in France'.[19] Holding stretches of what were no longer trenches, but scattered mud-filled shellholes, those battered units clung to their positions until fresh troops of the Canadian Corps captured what had been Passchendaele village and the fighting died away.

For many years there was much argument about casualty figures, but it has now become clear that those of the British Army numbered some 250,000 men, the Germans' much the same. The Green Howards took their share, if not more. The anodyne accounts of the fighting put together in H.C. Wylly's *The Green Howards in the Great War* do no more than hint at the effect of the battle upon its survivors; in 1926, its year of publication, individuals were trying to forget. The Official Historian, writing two decades later, was less inhibited:

> The casualties alone do not give the full picture . . . the discomfort of living conditions in the forward areas and the strain of fighting with indifferent success had overwrought and discouraged all ranks more than any other operation fought by British troops in the war, so that . . . discontent was general.[20]

Criticisms of Haig's handling of the battle were to be widespread. Many, the product of ill-informed hindsight, have been refuted. But there is no doubt at all that, as on the Somme, he pressed on with the offensive long after the collapse of the weather had made tactical success impossible. Nevertheless, the German Chief-of-Staff appears to have thought otherwise:

> The stubborness shown by the British bridged the crisis in France. The French Army gained time to restore itself and the German reserves were drawn to

Flanders. The casualties which Britain sustained in defence of the *Entente* were not in vain.[21]

★★★

Rumours had for some time been circulating in the 8th and 9th Green Howards that their 23rd Division might be transferred to another theatre of war. On 8 November these proved to be correct when the two battalions found themselves in trains bound for Italy, an extraordinary transformation from the misery of the Ypres Salient. Their formation was part of a force of six French and five British divisions sent south from the Western Front to rescue the Italian Army from defeat.

Although secretly allied to Austria-Hungary and Germany, fear of the Austrians combined with greed for territorial expansion in the Trentino and Trieste had instead persuaded Italy in 1915 to enter the war on the Allied side. The consequence had been disastrous. On eleven separate occasions the Italian Army, ill led and equipped and often half-starved, had assaulted the Austro-Hungarian positions above the River Isonzo on the eastern frontier. Each attack had been repulsed. Gallantly though the Italian peasants had usually fought, half a million of them were to die before the war ended. Eventually, in October 1917, an Austrian offensive, reinforced by German divisions released from the Eastern Front after the collapse of the Russian Empire, resulted in the rout of the hard-tried Italian defenders at the Battle of Caporetto. Despite their tribulations on the Western Front, France and Britain responded to the Italian Government's plea for help. But the attackers had moved too fast, outrunning their supplies. Before the Anglo-French force could enter the battle the Italians had managed to form a new and shorter defensive line along the River Piave, some eighty miles back.

The reaction in those two Green Howard battalions to this switch from the horrors of the Ypres Salient to the countryside of northern Italy, still barely touched by war, can be imagined. Only rarely did the dangers and discomforts of Italy approach those they had known in France and Belgium. They first entered the line on 2 December at the hinge of the new defensive line where the mountains met the Piave, their task to help hold the Montello, an isolated and commanding flat-topped hill. So ably did they and the rest of their brigade get down to improving the rudimentary defences that their corps commander directed all senior officers to visit their sector to see how sound defences should be constructed.[22]

Little happened in the Montello, however, except for some patrolling and shelling. Then at the end of March came a complete change of scenery. The 8th and 9th Battalions moved to the Asiago plateau, a natural basin in the Alps and an historic invasion route into Italy from the north, already the scene of much bitter fighting. There the Austro-Hungarians, again fighting alone, the German divisions having been withdrawn to the Western Front, launched in June their last and hopeless offensive. It had been aimed at taking the Piave defences in the rear, but their empire was on the verge of disintegration and their troops were sick of the war. The British divisions crushed the offensive so easily that the Green Howards, part of the divisional reserve, were barely involved.

More than half the Anglo-French force had already been returned to France in March 1918 to help stave off the German spring offensives. In September, with the reduction of infantry battalions in each division from thirteen to ten, the 9th Battalion was also to take train for France to join a reconstituted 25th Division, leaving the 8th to see some hard fighting in the Battle of Vittorio-Veneto, the final and victorious Allied offensive.

The battle opened with an assault crossing of the River Piave, a challenging obstacle, a maze of channels often fast-flowing and in places three miles wide. Faced with the problem of tackling Papadopoli Island, possibly the most difficult area of all, were the two remaining British divisions. Four days after the initial assault on 23 October, it became the turn of the 8th Battalion. After a night approach march on a compass bearing in pouring rain across a mass of narrow water channels and sand-banks, the Battalion succeeded in arriving safely at its forming-up point. The attack the next morning involved wading a 60-foot-wide fast-running tributary, in which many wounded men were swept away. For his work then and later, Sergeant W. McNally, MM, won his Regiment's twelfth and last Victoria Cross of the war, an extraordinary record for one with so few battalions. Except for the 12th Battalion, the pioneer unit of 40th Division, every unit of the Regiment that had fought in the war had gained the decoration.

The 8th secured its objective, but it had been bloody work which was to continue for two more days. During this time the Battalion pushed on into open country five miles beyond the river, fighting all the way. The cost was to be heavy, more than 120 all ranks either killed or wounded. Five days later, on 4 November, an armistice with the Austro-Hungarian Empire was signed. The war in Italy was over.

★★★

The dearth of manpower that had cut the number of battalions in infantry divisions had first hit the Regiment on the Western Front in February 1918 when both the 7th and the 10th Battalions were disbanded, two of the 141 that disappeared in France and Belgium. Just a few weeks before, that magnificent but sorely tried 7th Battalion had suffered a further severe loss, a harbinger of the future. Enemy aircraft had bombed its transport lines, killing eight men and forty-nine animals, and wounding a further twenty-four men. The air arm was now a factor in battle.

As 1917 drew to its end, the 13th Battalion saw something of what was to be a radical change in military tactics, the birth of massed armoured warfare. A British invention, the first few available tanks had been used in only small numbers for close infantry support on the Somme and at Third Ypres. By now, however, the new Tanks Corps was firmly in being. At Cambrai on 20 November, over almost undamaged ground, nearly 400 of those slow but still unreliable monsters, under an umbrella of 289 aircraft, led eight infantry divisions forward on a six-mile front. Forgoing any preliminary bombardment, surprise was complete. The Hindenburg Line was breached to a depth of some four miles. In London the church bells celebrated victory, but by the end of that first day, half the tanks were out of action and the crews of the rest utterly exhausted, as were the infantry. Worse, there was no reserve of tanks to exploit the success.

As the battle relapsed into the same old slogging match, on its third day the 40th Division, brought forward from reserve, was directed against Bourlon Wood, the key to the British left flank. Led by tanks, the 13th Green Howards, in the forefront of the attack, took their objective on the morning of 23 November, but at heavy cost. Their strength had been no more than 474 all ranks, half the usual number a battalion mustered for a major attack. When they were withdrawn three days later, less than 100 other ranks remained. For a week more Bourlon was the scene of furious fighting, before the Germans, who had somehow scraped together some reserves, struck back. Again it was stalemate.

★★★

This shortage of British troops for the Western Front had been accentuated by the proliferation of 'sideshows' elsewhere. Nearly half a million men had been engaged in the capture of Baghdad from the Turks; a third of a million were starting to push on beyond Jerusalem; a 600,000-strong Anglo-French force was involved in a futile campaign in Salonika; and there were also those five divisions, as well as a hundred heavy guns, doing little as yet in Italy. The influence of the 'Easterners' had gained the upper hand. In the French Army the situation was no better: their divisions now fought with only 5,000 infantry, half the 1914 figure.

German losses had been equally appalling, but Russia's collapse had allowed Hindenburg to switch a total of thirty-four divisions to the Western Front. In Picardy on 21 March 1918 the long-awaited blow hit the British Third and Fifth Armies. At dawn that day in dense fog, the infantry of sixty-two German divisions, led by picked units of storm troops and supported by the gas and high-explosive shells of 6,000 guns, began their assault over the old Somme battlefields. With horrifying speed the British forward defences were overrun.

In theory, mutually supporting posts sited in depth formed the framework of the British defences, but the shortage of both fighting troops and labour units had left many gaps. But some two miles behind those forward posts was a second line of redoubts, manned by reserve battalions. The story of the defence of one such redoubt by the 2nd Green Howards was told by its adjutant, the future Sir Herbert Read.

The first of the massed grey formations that approached about midday was mown down, but a second and equally costly attack that afternoon penetrated the forward Green Howard platoon positions. All that day the German assaults continued so that, by the following morning, Read remembered, 'beyond hope or despair', the survivors 'regarded all events with an indifference of weariness'.

> Again the morning was thickly misty. Our own artillery fire was desultory and useless. Under cover of the mist, the enemy massed in battle formation, and the third attack began about 7 a.m. We only heard a babel in the mist. Now our artillery was firing short among our men in the redoubt . . . This new

Above: *HRH Alexandra, Princess of Wales, painted by Lauchert and copied by William Dring. The original hangs in the breakfast room at Windsor Castle. The copy was presented to the Regiment in 1953 by the Boroughs of Beverley, Bridlington, Middlesbrough, Scarborough, Redcar and Richmond, all of whom have granted their Freedoms to the Regiment.*

Above: The 2nd Battalion holding the Menin Cross Roads in October 1914 by Fortunino
Matania. The figure in the foreground is Private Tandey who latter won the VC, the DCM and the
MM. The officer on the left with a map is Lieutenant-Colonel C. A. C. King who was shot by a
sniper later in the battle.

Below: After dinner in the mess of the 2nd Battalion at Bordon in 1889. Watercolour by W. Cutler.
It seems to have been a hive of activity, from whist to banjo-playing.

attack petered out . . . The fourth attack was delivered about midday . . . We fired liked maniacs. Every round of ammunition had been distributed. The Lewis guns jammed; rifle bolts grew stiff and unworkable with the expansion of heat.[23]

And so the attacks persisted all that second afternoon. The enemy had almost worked round to the rear of the redoubt and nothing could be heard of the units on either flank. With the Colonel (it was 'Nutty' Edwards, that veteran of the Tirah, Somaliland, the Boer War and of Arras) wounded, Read took over. Extricating the few exhausted survivors, he then commanded them brilliantly in a fighting withdrawal that lasted for seven more long days. Afterwards a DSO was added to his MC.

The two Green Howards likely to be remembered in the future for other than purely military achievements are Sir Herbert Read and Lord Edward Fitzgerald. Both were rebels against society; both are commemorated in the *Dictionary of National Biography*. Read, one of the most celebrated men of letters of his day, was for long after the war an anarchist, but in his later years accepted a knighthood for his services to literature. A North Riding farmer's son, he was to retain a deep affection for the Green Howards.

Within hours of the start of the German offensive, 50th Division was rushed forward from Fifth Army reserve to plug the gaps in the line. From then until the end of March the 4th and 5th Battalions fought an unending series of rearguard actions that cost them nearly all their officers and two-

Right: 2nd Lieutenant Herbert Read (later Sir Herbert Read, poet, essayist and art critic). He was to win both the MC and the DSO. (Mr. Thomas Read)

thirds or more of their other ranks. It was resistance of this calibre that in
the end brought the equally exhausted Germans to a standstill. Again each
side had suffered about 250,000 casualties – and in a mere sixteen days. In
this Picardy fighting, Lieutenant-Colonel O.C.S. Watson, who had been so
badly wounded in the Tirah as a subaltern, had won a VC, but again a
posthumous one, when commanding a battalion of the King's Own
Yorkshire Light Infantry.

Straight away Ludendorff scraped together fresh divisions for a further
offensive, this time in Flanders. As yet the few American divisions that had
arrived in Europe were being held back until they could be formed into a
separate United States Army. So once again the British and the French had
to withstand the German onslaught alone.

Already in the Ypres Salient were the remnant of the 2nd Battalion,
moved there to reorganize and recuperate. They were to escape the worst of
the fighting, but the two Territorial battalions were less fortunate. They had
been given but three days to rest and sort out their shattered companies
before being switched up to the Lys Valley. After absorbing a succession of
drafts, some as they moved, the majority either practically untrained boys or
old soldiers combed out of the back areas, they arrived just in time to take
the full force of the new offensive. In a further week of confused withdrawals,
mirroring the March fighting, each unit lost another 300 officers and men.
Two other Green Howard Battalions, the 12th and 13th, both of which had

*Below: Officers of the 2nd Battalion just after the German offensive of March
1918. On 30 April, 2nd Lieutenant J. S. G. Branscombe, sitting on the ground on
the right, was to be severely wounded. Despite his appearance, he was just twenty.
Most of the others would have been about the same age. The exception was the
45-year-old Quartermaster, Captain E. Pickard, on the left of the seated officers.
He was the only officer to serve throughout the war with the Battalion. The
adjutant, seated second from the right, is Herbert Read.*

been in the thick of the Somme fighting, fought on the left of these Territorials, the 12th Battalion as infantry, despite its pioneer battalion role. Again the Germans were held, but only just, this time with the help of French divisions rushed up from the south.

What little remained of 50th Division, its infantry numbering no more than fifty-five officers and 1,100 other ranks,[24] were then sent south to the Champagne country, to rest and absorb further vast drafts of the same type of raw recruit that had fought so well in Flanders. But once again they were unlucky. On 18 May a third German offensive fell upon them. In the Battle of the Aisne the 4th and 5th Green Howards finally perished.[25] There were no reinforcements left. Few units on the Western Front had, over the years, seen harder or more prolonged fighting.

By the end of May only a single Green Howard unit was still fighting in France, the 2nd Battalion, brought up to strength once again only when 6th Battalion was disbanded. The 12th and 13th suffered a similar fate, reduced to cadres to help train the Americans, now pouring into Europe at the rate of 250,000 men a month, but still largely dependent upon Britain and France for their guns, tanks and aircraft.

A final German offensive against the French in July brought their troops across the Marne, but there they were halted. Their last reserves had been destroyed, allowing the British and French in their turn to launch a series of crushing blows. From September onwards the Americans were able to play their part, but the heaviest burden was carried by Haig's battle-worn British Empire divisions.

In the forefront of this final British offensive was the 2nd Battalion Green Howards which won further distinction when it helped the Canadian Corps assault the Canal du Nord and turn the northern end of the Hindenburg Line. Back from Italy, the 9th Battalion had joined a reformed 25 Division. So it was that two units of the Green Howards were there at the end, fighting against picked German units still resisting with suicidal fervour even though the war was clearly lost. These final battles were to be as bloody as ever. In the last nine days of the war 175 officers and men of the 2nd Battalion were either killed or wounded. Another Green Howard, a regular soldier, Private Henry Tandey, who had already won both a MM and a DCM, was awarded a Victoria Cross while serving with 5th Battalion, the Duke of Wellington's Regiment. Since 8 August the British alone had lost 350,000 men.

The signing of the Armistice on 11 November produced in London and elsewhere scenes reminiscent of Mafeking. With the front-line units it was different. A member of the 2nd Battalion wrote, 'The officers and men took the news very calmly, it seemed too good to be true.'[26] They were still alive.

Ten Green Howard battalions had fought in Europe. Fourteen others, most of them reserve, training or labour units, had served at home or abroad. Of the 65,000 men who had enlisted in the Regiment, 7,500 had died; the wounded numbered 24,000. There were few homes in the North Riding of Yorkshire in which the words 'Green Howards' or 'Yorkshires' did not evoke tragic memories.

Left: Lieutenant C. N. Marshall, MC, serving with the 6th Battalion in North Russia in 1919.

Below: Light Car Patrol.

Bottom: Lieutenant A. C. L. Parry at ease in the back of a car that is either broken down or over-heated, but still bearing the defeated enemy insignia.

8. THE INTER-WAR YEARS: 1918-39

For some Green Howards the war was far from over. Two New Army units, the 6th and 13th Battalions, had been reconstituted at the end of July for service in, of all places, northern Russia.[1] The officers were mainly Green Howards, the rest a very mixed lot. Of the 422 reinforcements who joined the 6th Battalion during the last fortnight of July, fifty-seven only were Green Howards. The others had been culled from a score of different corps and regiments, a polyglot collection of English, Scots and Welsh.[2] But how did British troops become involved in the struggle between Whites and Reds in the Russian Civil War?

For most of the 1914-18 War the Germans had virtually closed the Baltic and Black Sea ports to Allied shipping, leaving Murmansk, near the Norwegian border, as the sole Russian winter lifeline to the west, because the Gulf Stream kept it largely free of ice. The collapse of Russia in 1917 posed new threats to the Allies, not least of which was the likelihood of German submarines being brought in sections by train to Murmansk, there to be reassembled and deployed against Allied convoys. In June 1918 a small, international expeditionary force, including British, Poles, French, Canadians, Italians, Finns, Serbs and Americans, under British command, invaded northern Russia and seized Murmansk. Far too many of these troops were either semi-trained recruits or battered old soldiers, and they had not been trained or equipped for Arctic warfare. In the event, their original role of retrieving munitions and equipment supplied to the Tsarist government had been widened, almost inadvertently, into support for the White Russians, a hopeless and invidious task which, even if successful, could hardly have led to Russia re-entering the war.

By the time the Green Howards arrived, the war against Germany was over and the Allies, with their main base at Archangel, were holding a vast frontage covering the White Sea and extending some 500 miles southwards. Although the ancient ship that carried their two Battalions had sailed on 17 November, a chapter of maritime accidents had delayed their arrival until 26 November.[3] Until February 1919 their time was to be filled with grinding guard duties, fatigues and train protection, the men working in temperatures down to minus 40 degrees centigrade. Daylight lasted for three hours. Battalions were split up into detached companies, but the 6th Battalion was detailed to form a mobile company, nearly 200 strong. Trained by Sir Ernest Shackleton, the Antarctic explorer, this company quite quickly took to snow-

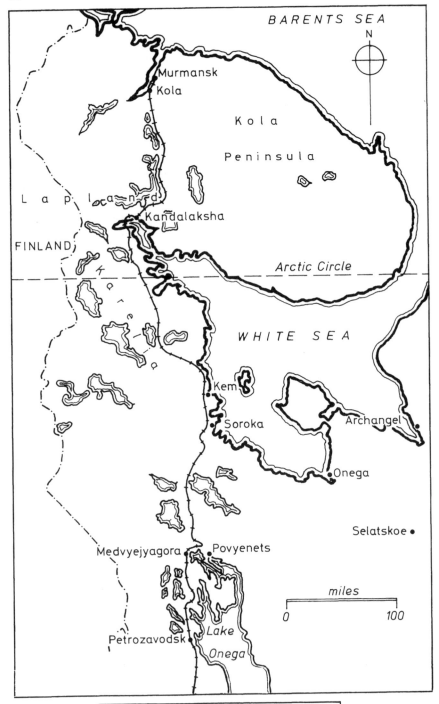

BARENTS SEA

N

Murmansk
Kola

K o l a

P e n i n s u l a

L a p l a n d

FINLAND

Kandalaksha

Arctic Circle

K a r e l i a

WHITE SEA

Kem

Soroka

Archangel

Onega

Selatskoe

Medvyejyagora   Povyenets

miles

0   100

Petrozavodsk   Lake

Onega

North Russia 1918-19

shoes and skis; baggage was carried on sleighs and winter clothing was provided similar to that used on polar expeditions.[4]

When the Bolshevik threat from the south against Murmansk slackened, Major-General Edmund Ironside, the Allied commander in North Russia, moved the two Green Howard units round to the Archangel area, on the far side of the White Sea, where the Red forces were far more active. Part of the 13th Battalion travelled by sea to Archangel in an ice-breaker; for the rest of the Green Howards it was to be an unusual journey. It began and finished by train, but the middle 400-mile section was by trotting horse- or reindeer-drawn sleighs, each carrying two men. At night shelter was found in the comfortable and solidly built peasant houses where the men, greatcoats often frozen solid, could thaw out. What was remarkable was the high standard of living then enjoyed by the *kulaks*.[5]

At the end of February the 13th Battalion reached the Seletskoe area, some seventy miles south of Archangel, in excellent shape, a fact commented upon by General Ironside who saw them arrive.[6] But on 26 February he received a horrifying telegram from Lieutenant-Colonel Lavie of the Durham Light Infantry, an able and experienced officer, who had taken over command of the 13th Battalion only three weeks before. His men had refused to parade.

General Ironside, who spoke more than a dozen languages fluently, including Russian, was a hulk of a man, frightening to confront. Afterwards he was to become CIGS. Beset with every conceivable problem from his hotch-potch international force, he had not anticipated mutiny among his British regiments. Travelling by sleigh, he caught up with the Battalion the following evening to find that Lavie had everything in hand. Lavie had gone straight to the billets and ordered the men to fall in without arms. This they did, but two ex-Army Pay Corps sergeants, neither of whom had been overseas before, had stepped forward and declared that the Battalion would do no more fighting. Sending a lance-corporal and a file to fetch their rifles, Lavie ordered them to escort the two sergeants to the guardroom. That ended the trouble and in due course the Battalion marched.

The two sergeants were tried by court-martial and sentenced to be shot, but Ironside had received secret orders direct from HM the King that no death sentences were to be executed. The men were given life imprisonment, but were released in 1931.[7]

Such trouble was hardly surprising. There had been a number of mutinies in France and England by both British and Canadian troops, impatient for their ineptly handled demobilization. How much worse it was for men who had experienced the miseries and dangers of that Russia winter, few of whom understood why they were still fighting when the war had ended months earlier. In any case, many disapproved of the British intervention. Nor did it help when the *Daily Express* insisted that 'The frozen plains of Russia were not worth the bones of a single British Grenadier'.[8]

Russian mud, which made much of the country all but impassable when the thaw arrived, was to receive the bones of many Green Howards during the coming months. To halt the steady Bolshevik pressure from the south towards Archangel involved holding the single railway line and the rivers, the sole arteries of communication with the rest of Russia. So it was that fighting devolved upon small forces. In this way, the Green Howards fought for the next three months, often in company detachments and switched from one area to another as the threat changed. More often than not they would be supporting and bolstering White Russian units which Ironside was equipping and training in the hope that the anti-Bolshevik forces might rally and form a stable government in North Russia, so allowing the British and Allied troops to leave. In defending and attacking small isolated posts, six officers of the 6th Battalion and thirty-six other ranks are mentioned as having been killed or wounded. The 13th was more fortunate, losing only three men. The Communists were not the only enemy. When summer came mosquitoes swarmed, bringing malaria as well as making life almost unbearable for those operating in the forests.

In June 1919 Ironside's British battalions, described in the official despatch as 'composed of low-category men selected originally as unsuitable for service in France and further severely tried by the rigour of an arctic winter'[9] were replaced by fresh troops, newly recruited youngsters. The Green Howards and the rest of those low-category men embarking at Archangel for demobilization, now tough and battle-hardened, were to be denied the special campaign medal which Ironside considered their due.[10]

With a spate of serious mutinies among the White Russian units emphasizing the futility of the campaign, all the Allied troops were soon afterwards withdrawn. The last had gone by October 1919, sailing before Archangel became ice-bound for yet another winter. Seldom had British troops fought in a more unpromising cause.

★★★

At the start of the Great War there had been fifty-two regular battalions of British infantry stationed in India and Burma. By the beginning of 1915 all but a dozen had left for more active work elsewhere, relieved first by Territorials and later by garrison battalions. Among that unfortunate dozen was the 1st Battalion the Green Howards. There in India it was to linger for the entire four and a half years of the war. As individuals and drafts were ordered away, its regulars were steadily eroded. One newly commissioned regular subaltern, arriving at a time when the fighting on the Somme was reaching its climax, was surprised to find himself once again repeating the drill and weapon training he had absorbed at Sandhurst, and listening to lectures on Marlborough's campaigns. The formal social life of India followed its age-old pattern, with 'calls' to be paid and cards left on the ladies of the station. This was the year when officers received permission to shave their upper lips, an indulgence gratefully received by some young subalterns whose obligatory moustaches were insufficiently hirsute.[11] The atmosphere of war, that subaltern complained, had hardly penetrated to India. As soon as he could he volunteered for the Royal Flying Corps.[12]

We have already heard of 'Nutty' Edwards. Then in charge of the 1st Battalion, in desperation he had penned a long and heart-rending appeal to his Colonel-in-Chief, HM Queen Alexandra, pleading that her Battalion 'suffering and eating our hearts out in silence' might be sent to France. Inevitably he received no more than an anodyne reply. When Edwards was next ordered to detail two officers for France, however, his own name and that of his adjutant in some way went forward and to France they went, Edwards eventually to command a brigade.[13]

Not until the war was over did the 1st Battalion see action. Stationed for most of those years in or near Rawalpindi, on 5 May 1919 it mobilized for what was to be known as the Third Afghan War; at a little over 500 strong, the Battalion numbered less than half its proper establishment. That month the Amir of Afghanistan, who had succeeded a murdered father, declared *jihad* or holy war, primarily to conciliate the generals to whom he owed his throne. His moment was well chosen. British troops were impatiently awaiting their overdue demobilization; the Indian Army was war weary after the despatch of a million men overseas. Unrest pervaded the rear area administrative troops and there was serious rioting in the Punjab.[14]

The Amir's error was to launch his three-pronged invasion through the passes towards India with troops better suited for guerrilla than formal warfare. Nevertheless it was to be one of the most difficult campaigns the Raj had yet fought on the Frontier, one that engaged 750,000 men and nearly 500,000 animals. Those Green Howards, most of them wartime soldiers, were to undergo exactly the same hardships their regular predecessors had known in the past among those rocky precipitous slopes – intense summer heat, choking dust, flies, fatigue and foul food. The sole change since the Tirah was the presence of some obsolete aircraft of the Royal Air Force (successor to the Royal Flying Corps), invaluable for searching out the enemy and hitting them with diminutive bombs.

Brigaded with three Gurkha units, the Battalion had a lance-corporal Lewis-gunner killed and two soldiers wounded by snipers as it toiled up the narrow defiles of the Khyber Pass; the local Afridis, joining in to help their co-religionists, had been granted an almost clear run by their fellow tribesmen of the Khyber Rifles whose job it should have been to protect the route. Landi Kotal Fort, at the summit of the Pass, was reached on 12 May. By then the Afghan incursion, penetrating to within two miles of the Fort, had been repulsed, and the Green Howards joined the force that was pushed westwards down the Pass to debouch into the more open country beyond. In open fighting that involved attacking across bare ground with inadequate fire support, it helped route the Afghan Army, losing just five men in the process. There it then stayed for the next four months in sweltering heat that averaged 117 degrees Fahrenheit in the shade.

Although Jelalabad could have been taken, the error of advancing farther into Afghanistan was avoided, largely because of the near impossibility of guarding the lines of communication. In any case it was unnecessary because in June the Amir began to make overtures for a peaceful settlement, a request which the Indian Government readily granted. It was the *status quo* again, as it had been after the two previous Afghan Wars, something that the Soviet Union would experience more than sixty years later, after their abortive incursion into that forbidding country.

But the Green Howards had more still to do. The local Mohmand tribesmen now joined in, attacking the piquets sited to protect the British camp. In a bloody little struggle to occupy one of these summits, a subaltern and four other ranks were killed and nine men wounded.

In September the Battalion marched back into India. It had been another sideshow, but one, unlike the North Russia campaign, almost unnoticed except by the participants. In all, nine hundred British and Indian troops had died of disease, two-thirds from cholera. The killed numbered 236 and the wounded 615. The scale of the fighting was nothing compared with what had happened elsewhere. Nevertheless, like those others in the 6th and 13th Battalions, awaiting demobilization and caught up in a war that did not seem to concern them, the 1st Battalion men had done all that had been asked of them. Only thirteen of the thirty-three officers who served with it on the Frontier had been Green Howards; in the ranks the proportion was much the same.

★★★

In July 1919 the 2nd Battalion's cadre was back from a month's leave, after having been given a civic reception in Richmond on its return home from France. Its strength numbered four officers and forty-three other ranks, eighteen of them bandsmen; by the autumn, once again a battalion, it was deployed under active service conditions in Ireland.

The second half of the 19th century had seen the growth of the Fenian secret society. Its aim was to establish an Irish Republic by force of arms, its boast that its members were the first revolutionaries in the world to use dynamite. Parallel with the growth of this violent body was the political pressure for Home Rule, the introduction of which in 1914 was prevented only by the outbreak of war. But, as ever, England's difficulties were Ireland's opportunity. The delay provided the excuse for violent action, and the subsequent ill-planned and incompetently executed 1916 Easter Week Rising in Dublin resulted in the death or wounding of 469 soldiers and forty-three police officers.[15] Afterwards sixteen rebel leaders were executed: the Catholic Irish had the martyrs they sought.

Although the country seemed quiet throughout the remainder of the Great War, the Nationalist groups were making their preparations. Impetus was provided by the fear that conscription would be extended to Ireland, a measure for which the Ulstermen had long been agitating. Southern Irishmen had enlisted in their tens of thousands to fight against Germany, but this hatred for conscription (never to be introduced) rallied the many Irish who had viewed the Easter Rising either with apathy or distaste.

The difficulty of reconciling the interests of the North delayed the implemention of the promised Home Rule. So it was that violence spread quickly after two policemen were murdered in Tipperary in July 1919. It was to Tipperary that the 2nd Battalion was sent.

At first this violence was directed against the scattered police posts, manned sometimes by as few as four members of the Royal Irish Constabulary. Smaller stations quickly became untenable, the larger ones fortresses, so that many rural areas were virtually abandoned to rebel control. With the assassination and terrorization of police officers, intelligence became almost unobtainable and proper co-operation between police and army rare. Details are limited of how the Green Howards were employed during these early months of the insurrection,[16] but guard duties seem to have absorbed the greater part of their small strength, leaving little time for training and improving the physical fitness of the new recruits, most of them ill nourished and under educated because of the war. In April 1920 the Battalion was one of the weakest of twenty-one infantry units serving in the South.[17]

In June 1920 the Green Howards moved from Tipperary to Rathkeale, where Calladine's company had been stationed nearly a hundred years before at the height of the Whiteboy disturbances. Described by him then as 'only a poorly built and straggling town',[18] it had not much improved. Straight away the Green Howards were deployed in small detachments around West Limerick, their work far more unpleasant than in Tipperary, as is clear from the comment in the *Gazette* that 'We hate our job here, but none of us would like to leave Ireland without first seeing an end to Sinn Fein and its despicable tale of murder, robbery, intimidation and anarchy'.[19] A little later, however, they were reporting 'enormous success' with signs of weakening activity by the 'Rebels', as they were known, and an improvement in relations with the local people.[20] Then, at the year's end, they were switched back to their old area in Tipperary and their work became more serious. Because of this increasing Rebel violence and also of civil unrest in England, the aftermath of the war, all ranks were recalled from leave and courses.

The Green Howards' move to Tipperary followed the murder of a dozen British officers, most of them engaged on intelligence work, in their Dublin billets on 21 November, a day to go down in history as 'Bloody Sunday'. There followed a wave of arrests, the restoration of internment without trial and the placing of the four south-west counties – Tipperary, Limerick, Cork and Killarney under martial law.

Although the Green Howards may have managed to establish friendly relations with some of the local people, Irish distaste for the Crown forces had been further aggravated by the atrocities of the 'Black and Tans', the non-Irish ex-soldiers brought in to augment the cowed Royal Irish Constabulary. Even more of a byword for ruthless behaviour were the hated Auxiliaries, a gendarmerie of unemployed wartime officers. Nor, sad to say, was indiscipline confined to these newly recruited policemen. A Green Howard bandsmen remembered how 'swift and devastating' reprisals followed the ambush and death of some soldiers of another battalion nearby. Armed with entrenching tools, men retaliated by smashing the windows of almost every shop in Tipperary's main street and looting the contents.[21] It was at about this time that the Green Howards were further embittered by the news that a popular and brilliant young regular officer, who had been promoted into another regiment during the war, had been kidnapped and then shot after a farce of a 'court-martial'.[22]

Indiscipline was hardly surprising. Demobilization and drafts for India continued to strip away the few experienced privates and NCOs whose experience, in any case, was limited to the trenches of Flanders, and who now faced the physical and psychological challenges of guerrilla warfare. Through lack of time, standards of training were abysmal: the bandsman mentioned earlier never fired a rifle course until two years after he left Ireland.[23] There was also a shortage of every type of equipment, including transport and signals stores.[24] Nor was morale especially high. An entry in 2nd Battalion's Notes in the *Gazette* went so far as to describe the Green Howards as having arrived in Ireland two years earlier 'even then war-weary'.[25]

Despite all this, the Battalion in the end gave an excellent account of itself. Easter was quiet in Tipperary except for unsuccessful Rebel attacks on billets and sentry posts in the various towns and villages around which companies and platoons were scattered.[26] Then, at last, one of the countless patrols was rewarded with success.

Over bold, the Irish Republican Army had begun to operate in 'flying columns', usually about twenty-five strong, billeted on friendly farms and moving rapidly in rubber-wheeled pony traps or on bicycles to mount its ambushes. The supply and movement of the Crown forces became even more hazardous but these new Rebel tactics produced similar counter-measures. Army patrols of up to three platoons, often composed of volunteers, moving on foot or bicycle and wearing rubber-soled boots, sometimes shorts, would cover up to twenty-five miles daily. Morale rose in the fine summer weather, and the army was soon a force to be reckoned with.[27]

On 2 May 1921 one such Green Howard bicycle patrol consisting of a subaltern and ten men, who were accompanying three policemen, ran into a large ambush, but extricated themselves with only a single man slightly wounded. In the exchange of fire at least five Rebels were killed and a number wounded. Placing the dead on farm carts, the patrol continued on its way, only to be twice attacked in a five-hours' running engagement against some 200 Rebels. In all, ten members of the IRA were killed and thirty-five wounded; the Green Howards lost only one more soldier seriously wounded. It was described in the *Gazette* as 'the finest show in Ireland up to date'. The previous day another patrol had gained a similar success, killing two, wounding one and capturing four Rebels, all armed. Congratulations poured in to Battalion HQ.[28]

Although actions such as these made clear to the IRA the dangers of working in formed bodies,[29] from May onwards casualties to the Crown forces mounted and sabotage increased. In Tipperary an unarmed Green Howard unwisely taking a solitary afternoon walk was shot three times in the hand (so that 'he could never use of a rifle again')[30] before being 'knee-capped'; he was lucky to escape with his life. Three officers belonging to other units in the garrison were less fortunate. Murders of police and civilians continued unabated, as did attacks on isolated detachments of the Regiment.

But the rebels were losing heart and political pressure was mounting on the British Government to find a way to halt the unhappy conflict. As ever,

Above: Funeral of Captain Henderson, murdered by the IRA and buried at Aldershot on 14 December 1923.

the problem was to reconcile the Protestant North to Home Rule, but on 11 July 1921 a truce came into being, to be followed on 21 December by the Anglo-Irish Treaty. The South gained dominion status as the Irish Free State, while the North accepted a measure of self-government within the United Kingdom. For the more extreme Republicans such a compromise was unacceptable, and a murderous civil war was to rack the country until they were crushed by their fellow-Southerners in 1923.

The Treaty marked the departure of the larger part of the British garrison, only a few troops staying to guard three naval bases retained by the British Government for use in case of war. To Queenstown went the Green Howards, and near Cork on 27 April 1922, Lieutenant Robert Henderson, MC, the Battalion Intelligence Officer, was kidnapped with two other officers and their driver and afterwards murdered. Months later their bodies were recovered, but why the killings occurred after the shooting had officially ceased has yet to be revealed.[31] It was a tragic epilogue to a sad interlude.

Later that year, the 2nd Battalion, now an experienced and efficient unit, moved north to Belfast.

★★★

Army Order 509 of 1920 contained splendid news for the Regiment. What had been for 178 years no more than its treasured nickname was to become its official designation. No longer would it be confused with the several other regiments that carried 'Yorkshire' in their titles. In the Army List of 1920 'Alexandra, Princess of Wales's Own Yorkshire Regiment' became 'The Green Howards (Alexandra, Princess of Wales's Own Yorkshire Regiment)', shortened to 'The Green Howards' as its previous title had been abbreviated to 'The Yorkshire Regiment'.

Primarily responsible for the new title had been Lieutenant-General Sir Edward Bulfin, a fine fighting soldier who had commanded in action and with great distinction everything from a brigade to a corps. When the death

*Above: First post-war Regimental dinner on 27 May 1920 at the Princes Hotel,
London. Far side: Lt A. W. Hawkins (nearest camera), Capt B. C. W. Williams,
Maj M. L. Ferrar (the first regimental historian and founder of the Museum), Maj
M. H. Tomlin, Wing Commander A. L. Godman, Brig G. Christian, Lt Col
E. M. Esson, Lt Col N. E. Swan, Maj H. C. Cumberbatch. Near side: Maj B.
Cuff, Lt Col M. D. Carey, Col. T. W. Stansfeld, Lt Col C. T. Hennah, Lt Gen
Sir Edward Bulfin, Lt Col G. B. Mairis, Lt Col R D'A Fife, Lt Col H. F. Lea,
Major H. Levin.*

of Lieutenant-General Sir Edmund Franklyn in October 1914 deprived the
Army at that most critical juncture of one of its most able army commanders,
Bulfin had succeeded him as Colonel of the Regiment. He afterwards
confided to a close Regimental friend that the change of title had failed to
find favour in one especially important quarter. King George V had
subsequently sent for him to tell him that he had insulted his mother, the
Dowager Queen Alexandra, by recommending the change, adding rather
inaccurately that 'you are the only regiment in the Army to be called after a
commoner' (The Duke of Wellington's Regiment had apparently slipped His
Majesty's memory).[32]

★★★

The immediate post-war years were strenuous indeed for the Army,
stretched as it was by campaigns in Afghanistan, North Russia, Ireland and
several trouble-spots around the world. Soon, however, it was to settle down
to its old role of Empire policing, half its battalions overseas with the rest at
home, often at little more than cadre strength, supplying annual drafts to
those abroad.

In an atmosphere of naïve pacifism, inevitable in such a war-weary
country, the taxpayers could perhaps just manage to find some virtue in
providing money for the Royal Navy and the Royal Air Force, their ultimate
safeguard against starvation and bombing. But it was hard to see the reason
for paying for an Army with no obvious enemy to fight. The so-called and

invidious 'Ten Year Rule', the premise that no major war would take place for ten years, an assumption to be extended annually, would by itself have effectively ensured that the Services would be starved of money; the economic slump of the twenties and thirties made it certain.

There was no money to pay for the development of new Army equipment. It was hard to persuade even the unemployed to enlist for two shillings daily, enough after stoppages for only two pints of beer and a packet of cigarettes. With the pay of a subaltern on home service just covering his mess bill and promotion slow, private means were still a near necessity. In 1924 the senior Green Howard captain of twenty years' service could see no prospects of a majority;[33] cheap sport with the leisure to enjoy it was almost the sole attraction of an officer's life.[34]

Despite these apparently unsurmountable difficulties, those two Regular Green Howard battalions still produced high standards in their limited sphere of shooting, drill, physical fitness and minor tactics. Nor would it have been easy to better its long-service corps of fine warrant officers and NCOs.

Right: Guard mounted by the 1st Battalion at Aldershot on 9 July 1930 for HIH Prince Takamatsu of Japan. The guard commander, facing the Prince, is Captain A. E. Robinson, later Colonel of the Regiment. The Colour Ensign is 2nd Lieutenant S. M. Boyle, who was to be killed on 17 June 1944.

Right: The 1st Battalion halts for an hour while on the march from Aldershot to Tidworth in 1930. Main roads were still relatively traffic-free.

***

After eighteen years abroad 1st Battalion arrived back in England in 1927. On coming down from the Frontier in the autumn of 1919, it had been rushed to Palestine, reinforcements for the Middle East to cope with the troublesome aftermath of Turkey's defeat and the break-up of the Ottoman Empire. The Easter of 1920 gave it a first taste of tackling Jewish-Arab riots,[35] a thankless task resulting from Britain's new Mandate over Palestine conferred by the newly created League of Nations.

Then, after no more than a year's stay in the Holy Land, the Battalion returned once again to India for a further five years before staging for a short time in Egypt on its way home. The Battalion's stay in England was to be short. On 2 April 1927, having been brought up to strength by reservists from five other Yorkshire regiments,[36] it was being ferried ashore at Shanghai on a cold and misty morning past a vast array of warships of all nations, its first experience of the Far East.

The cause of this sudden move was China's unsettled state. Throughout the twenties the country was beset by warring armies. Chief among these were the Nationalists, supported by the Soviet Union with military advice and munitions. Advancing northwards in 1926, these Nationalists clashed with their rivals in the area of the River Yangtse, so threatening the safety of the international and French Settlements in Shanghai. The presence of the foreign-controlled Settlements, set up in the middle of the nineteenth century and among one of China's main trading outlets abroad, was understandably loathed by Chinese of all political complexions.

When their nationals higher up the river at Hankow and Nanking were attacked, forces from the USA, the United Kingdom, Japan and France were directed to Shanghai. Britain's share was three brigades of infantry, one Indian and two from the United Kingdom. To help cope with the influx of refugees and keep out the warring factions, the Green Howards had the task of manning a 2,500-yard sector of the twenty-four miles of barbed wire entanglement around the International Settlement. In sandbagged and sometimes concreted pillboxes, they were faced by very young and shabbily dressed Chinese Nationalist troops, armed with usually rusty rifles.

There was some sniping. The subaltern who had led that successful patrol in Tipperary five years before was hit in the neck by a brick, but no one else was hurt. Moved into reserve, the Battalion trained in street-fighting, a much-needed skill but one that the Army had had little opportunity to practise.

***

For the 1st Battalion Shanghai had been an interesting but uncomfortable interlude, brought to an end when it sailed for home in January 1928 by way of the Suez Canal. There at Port Said on 2 February it met its 2nd Battalion, the first occasion since 1889 that the paths of the two regular units had crossed.

The 2nd Battalion had arrived in Egypt the previous year after a tour in Jamaica and Bermuda, having left Dover in 1925 at the start of what was

Above: 1st Battalion Transport in Shanghai, 1927.

to be a quarter-century of peregrination around the world. When, in the summer of 1929, it was beginning to think about its onward journey, oddly enough to Shanghai where a small permanent garrison was now stationed, it was ordered on 24 August 1929 to mobilize for Palestine. Communal trouble had once again erupted there, the most serious to date.

The root of the problem was the Balfour Declaration of 1917, the product of idealism and the need to influence American-Jewish opinion on behalf of the Allies. British sponsorship of 'a national home for the Jewish people' was promised, a pledge that unhappily conflicted with others to the Arabs encouraging them to revolt against the Turks. With the war ended, increased Jewish immigration caused the Arabs to fear for their future. The seeds of conflict – nationalistic, economic and religous – were myriad.

In the immediate post-war years, misplaced faith abounded in the ability of the new air weapon to cope cheaply with every defence problem. One result was that internal security in much of the newly acquired territory in the Middle East was handed over to Air Ministry control. All army troops were removed from Palestine, to be replaced by a few Royal Air Force armoured cars. The consequence was that the Green Howards were one of three infantry battalions to be rushed post-haste to assist the small and exhausted police force, almost entirely locally recruited, when serious rioting in Jerusalem was accompanied by widespread massacres of the Jews and attacks on their *kibbutzim*.

The first Green Howard company to arrive by train on 25 August was despatched immediately to Haifa.[37] Arriving there at 7 a.m. next day, by midday they had all under control and the Arab mobs disarmed and dispersed, even though two platoons had been separately to deal with fresh troubles outside the city.

In this way the Battalion worked during the following weeks, responsible for the security of the whole of northern Palestine. Switched from one area to another as the need arose, they dealt impartially with both sides, patrolling narrow streets, rescuing parties of Jewish women and children, and chasing Arab gangs over the tortuous hills in the heat of high summer. On one occasion a Green Howard NCO coolly outfaced two rival parties drawn up in battle array near Nazareth by marching his single section between them. Often under fire, and using their own weapons freely when need be, they experienced internal security work at its worst, supporting a police force too weak to cope in a country to which they were strangers. By the second week in September all was reasonably quiet again, but it was fortunate that the threat from thousands of Bedouin tribesmen from across the River Jordan had failed to materialize, possibly the result of vigorous patrolling of the Jordan Valley by Green Howards working with armoured cars of the 12th Lancers, one of the first of two cavalry regiments to exchange its horses for machines.

But the British Government had learned its lesson. In future two infantry battalions were to be stationed in Palestine.

★★★

After this short but active interlude in Palestine it was the 2nd Battalion's turn to serve in Shanghai, on this occasion a planned move. There it stayed until the following year, its tour enlivened for all ranks by anti-piracy guards on ships trading along the River Yangtze. During this often dangerous work Lance-Sergeant T.E. Alder showed fine leadership and bravery when his four-man patrol rescued one such vessel under heavy fire from both river banks. For this he was awarded the Empire Gallantry Medal, the forerunner of the George Cross to which all EGMs were converted in 1940. His was the first of three such decorations to be won by members of the Regiment.[38]

From Shanghai in 1930 the 2nd Battalion moved once again to India. There life followed its age-old pattern. For the officer there was, as ever, sport and generous leave to enjoy it. In their warehouse-like barrack blocks, the efforts of the *punkah-wallahs* toiling throughout the night did give the soldier some small relief from the sweaty heat of the Plains, before the barber

Right: A camp picquet at Ghariam.

Right: On the march between Bannu and Waziristan. Lieutenant-Colonel T. A. Peddie leads his battalion accompanied by his adjutant, Captain W. O. Walton.

arrived to shave his sleeping face and the *char-wallah* brought his reveille tea.

This tedium, with its accompanying prickly-heat, was relieved on September 1937 by a foray to the North-West Frontier, that first taste of active service so avidly anticipated by the unmarried young of all ranks. The Battalion's tour among those inhospitable hills was to be the best recorded of all the Regiment's sojourns in the old Army's finest individual and small unit training area.[39]

A year in Razmak at the start of 1938 had been planned, but a temporary shortage of troops brought forward the move. Yet, further fighting had erupted the previous summer, just another attempt by the very independent Pathan tribesmen to rid themselves of the infidel who discouraged them from raiding into the Plains of British India or exacting tribute from the caravans that traversed their mountain passes. That autumns's fighting allowed the Battalion first to learn and then perfect those ancient Frontier skills of picqueting the heights to allow free movement along the valley floor. Fitness and boldness by junior leaders were the prerequisites for success, and in these the Battalion quickly excelled. The cost was no more than three men wounded.

Just before Christmas, when thick snow brought a seasonal end to hostilities, the Green Howards moved from their tented camps into Razmak, a vast fortified cantonment built after the 1920 fighting, the seventeen campaigns fought since 1852 having persuaded the Government of India of the need to occupy central Waziristan. Razmak was the answer, reached by a road that was suitable for motorized transport and which ran both north and south back to the Plains. Surrounded by a 5-foot-high stone wall, six battalions and supporting troops lived in permanent huts, complete with messes, canteens, workshops, tennis courts, a cinema and even a roller-skating rink. But there was not a single woman. From Razmak, columns, sometimes two brigades strong, could sally forth to discipline especially recalcitrant tribes. Except of course for the higher degree of comfort and the help of a few lightly armoured vehicles, life was much the same as it had been in 1920, if not 1897.

The work was hard, especially when road-making in the summer heat. It could also be dangerous. During one of the routine, so-called 'columns', expeditions that often lasted for three weeks or so, a Green Howard captain and four private soldiers lost their lives and a number of other men were wounded. Even in Razmak, sniping from the surrounding hills was not infrequent and picquets were needed to protect the firing-ranges and football grounds that lay outside the perimeter. Life was rarely dull.

When the 2nd Battalion moved down to Ferozepore in the Punjab after its year's tour, higher commanders were loud in their praise of the cheerfulness and 'go' of its Yorkshiremen. Fighting as they were alongside Gurkha and Frontier Force Rifle battalions, units with far greater experience of the Frontier, their professional skills well honed, these were compliments indeed.

***

Such was the shortage of infantry in 1937 that both regular Battalions of the Regiment were once again overseas at the same time, both on active service. It came about through the renewal of violence in Palestine, not just the isolated rioting and murders of 1929 but what amounted to a full-scale

Left: Officers of the 2nd Battalion, June 1938. The climate on the Frontier was cool enough for them to be able to wear the 'blue patrol' uniform, introduced again for ceremonial purposes for all ranks for a time after the coming war. Back row from left to right: Lt C. S. Scrope, Lt G. Ritchie, 2/Lt W. K. Pryke, 2/Lt J. M. Forbes, Lt. M. W. T. Roberts, Lt S. M. Boyle, Capt C. E. W. Holdsworth, 2/Lt J. F. Atkinson, Lt. J. G. Middleditch, 2/Lt E. G. P. Harrison. Middle row: Capt F. E. A. Macdonnel, Capt W. E. Bush, Maj H. N. Bright, Capt W. O. Walton, Lt-Col T. A. Peddie, Maj C. E. Brockhurst, Lt W. J. Lipscombe, Capt F. C. Ainley, Capt J. S. G. Branscombe. Front row: 2/Lt Bewoor (Indian Army attached), 2/Lt P. D. H. Fox, 2/Lt Khan (Indian Army attached).

Arab rising aimed at halting, if not reversing, the continuing Jewish immigration. Faced with the impossible task of finding a compromise between Arab and Jewish aspirations, all that the British could do was try to keep the two races from each other's throats and to protect themselves. For this twenty-three battalions and an armoured car regiment had to be gathered to hold down the Arabs, a thankless task indeed. Less than a decade later the British were to be attacked by the Jews.

Moving from England in 1937, the 1st Battalion reached Palestine in October 1938 by way of Malta, a delightful place to be stationed and as always one enjoyed by all ranks. For most of its short five-months' stay in the Holy Land, it was to be based on Nablus in the Judean Hills. Their commander there was to be Brigadier H.C. Harrison, lately their colonel. Known throughout the services as 'Dreadnought' and possessor of several England rugger 'caps', he had transferred to the Army from the Royal Marines.

By the time the Green Howards arrived large areas of the countryside were under rebel control: road movement was dangerous, policemen and administrators were being murdered. The Battalion had two tasks:[40] first to keep open the main road running from Jerusalem to Haifa by way of Nablus; second to deal with the rebel gangs living in the hills. To do this it was organized into three motorized flying columns, each of two platoons, a 3-inch mortar section, a 'wireless telegraphy set' and a rearguard or escort section. The donkeys which carried the mortars and radio when the column left its transport to move across country also had their trucks. The Battalion's fourth company was, as ever, needed for guards and escorts.

As always, internecine feuding was rife among the Arabs, so making the gathering of intelligence not too impossible a task and brought an early success. A couple of weeks after the Battalion arrived, news came that a gang was collecting food from the village of Beit Furuk. Straight away, Gloster Gladiator fighters of the RAF cordoned the village, shooting up anything that moved after dropping warning leaflets.

At the same time, a Green Howard column was alerted. The first five of its ten-mile journey from Nablus was covered in its 15-cwt trucks, the rest on its feet. Then, as its platoons began to surround the village, a Gloster flew low over them and dropped a message asking them to move quickly to a shot-down fighter. As a section was despatched towards the smoking remains, parties of armed Arabs were seen trying to escape through the olive groves. In the exchanges of fire that followed during the next three hours, about a dozen rebels were killed. A single Green Howard was slightly wounded; the pilot was later to die. With continued Arab sniping from the surrounding hills as the column began to pull back towards the road, a newly arrived detachment of mountain artillery did useful work in helping to cover the withdrawal.

This action was noted as having been typical of many others, successful and unsuccessful, fought during the succeeding months.[41] More often than not, however, the Arabs had flown by the time the troops arrived, but the subsequent searches rarely failed to uncover quantities of ammunition and often weapons as well, at times in sizable quantities. It was to be a busy six months. The Arabs made unsuccessful attempts to bomb and snipe company billets; the risk of mines rendered road travel a chancy business; there were few days in the month when either one or both columns were not out. Nights in bed became a luxury.

The Battalion weathered the tour with no more than one lance-corporal killed and four other NCOs and the signals officer wounded. The campaign was, however, marked by more men being hit by rounds accidently discharged by their own side than by enemy action, a far from uncommon accompaniment to internal security operations. This did not happen to the Green Howards, but a ghastly tragedy occured just a few weeks before they left the country when an exploding petrol tank in a vehicle fire caused the loss of thirty-two men, three of whom died. For their courage during this disaster Corporal Atkinson and Private McAvoy were awarded the Empire Gallantry Medal. Like Lance-Sergeant Alder's they were in 1940 converted to George Crosses, bringing up to three the total won by the Regiment.

Two subalterns won the Military Cross and Lance-Corporal Peacock, later a redoubtable RSM, was awarded the Military Medal. Certificates for Distinguished Service in Action were presented to them – and several others – by the Divisional Commander, a certain up-and-coming Major-General B.L. Montgomery, who had formed an admiration for the Battalion when it was at Plymouth in his brigade.

After a last month in new surroundings along the Syrian frontier, the 1st Battalion embarked for England in April 1939, its job well done. Military pressure had done much to bring that Arab rebellion to an end. Quite soon attention was to be diverted to greater events elsewhwere.

When at Catterick that July, HRH The Princess Royal, sister to the newly crowned King George VI, presented new Colours to the Battalion, an occasion that marked also the 250th anniversary of the Regiment's raising, celebrated a year late, nearly everyone was wearing the General Service Medal with the clasp 'Palestine', so matching that India General Service Medal awarded to those other Green Howards, down from Razmak.

9. SECOND WORLD WAR: NORWAY TO TUNISIA, 1939-43

As the 1st and 2nd Green Howards were toiling over the hills of Palestine and the North-West Frontier, a major world conflagration was becoming increasingly imminent. Ever since Hitler had attained power in 1933, the spectre of a resurgent and aggressive Germany had haunted Europe. By the mid-thirties even the Treasury was persuaded to loosen its purse-strings, and a start was made on remedying the gross deficiencies in the British armed forces – a very tentative start in the case of the Army. Because the bomber was assumed to be near invincible, priority was given to equipping the RAF, first with bombers for reciprocal attacks on Germany and then fighters to defend the home base. The little extra cash that could be spared for the Army was spent largely on anti-aircraft artillery to support the fighter defences. Otherwise progress was slow indeed, especially in the production of armoured vehicles. Few of the country's leaders were ready to face up to the likelihood of British troops once again taking their place alongside the French on a European battlefield.

However, the infantry began to receive some new weapons such as the much needed 3-inch mortar; Bren light machine-guns, a few light armoured carriers and the all but useless Boyes anti-tank rifles also arrived. More radios became available and, in the teeth of much opposition, mechanization was making some progress: except in some Yeomanry units, the internal combustion engine at last replaced the horse (even though in the German Army much of the transport in the infantry divisions was still horse-drawn, at the war's end).

Something also had to be done to make the profession of arms rather less unattractive. In 1936 all subalterns with thirteen years' service had been given their captaincy. Two years later, the controversial and reforming Secretary-of-State for War, Leslie Hore-Belisha, introduced time promotion, eight years to captain and seventeen to major: eight Green Howard captains thereby benefited. Some improvements were also made to the pay, marriage allowances and terms of service of the other ranks, and a programme of barrack building was begun.

But none of this was enough to attract the numbers of men needed – in August 1939, the 1st Battalion was so under strength during formation training at Catterick that sections were still being represented by flags. So it was that when Hitler finally swallowed the rump of Czechoslovakia in March 1939, Britain and France having acquiesced in that luckless country's dismemberment at Munich in the previous September, a measure of conscription was introduced into the United Kingdom, the first in the country's history in peacetime. The period was for only six months and the anodyne term 'militiamen' was used, a euphemism for 'conscript', so unpalatable to the British public. In June the first of those Militiamen, dressed in the new and shapeless battledress that was to disfigure the British Army for the next two decades, reported for training at Richmond.

During the post-war years of neglect, a handful of enthusiasts had kept alive the 4th and 5th Battalions, based upon Middlesbrough and Scarborough respectively, part of the Territorial Army, reconstituted as such in 1921. In these two Battalions, as in the rest of the TA, the September Czech crisis had produced a surge in recruiting, volunteers comparable in quality and in quantity to the men who had flocked into Kitchener's Armies twenty-five years before. When, in March 1939, Hore-Belisha impulsively doubled the Territorial Army from thirteen to twenty-six divisions (without even referring the matter to the General Staff), the Green Howards' two new Battalions, the 6th and the 7th, were both quickly at full strength.

Numbers were one thing, equipment and leaders another. At first the regular element of the Regiment could not find even adjutants for these two new units. Rifles, uniform and enthusiasm were all that they had. But this was hardly surprising in view of the widespread shortages.

***

Ignoring with contempt the guarantees Britain and France had made to Poland in March 1939, Hitler invaded that country on 1 September. Two days later a stunned nation listened on their wireless sets to their Prime Minister, Neville Chamberlain, telling them that Britain had declared war upon Germany in accordance with its treaty obligations. That afternoon the French also were at war.

On the evening of 4 October the Green Howards of the 1st Battalion entrained at Richmond, as so many of their predecessors had done before them, bound for France as part of 5th Infantry Division, its commander their own Colonel, the then Major-General H.E. Franklyn. They were a magnificent body of officers and men, well trained but ill equipped for modern war.

In January 1940 the 4th and 5th Battalions followed them. They still belonged to the famous 50th (Northumbrian) Division, its badge the initials 'TT' standing for the Rivers Tyne and Tees. Organized as a mechanized division a few months before, with its infantry all truck-borne, again it was one of the first Territorial divisions to see action. During those winter months, the division had trained hard among the Cotswold hills and left behind them happy memories of friendly, well-behaved soldiery with all but impenetrable accents. Three months later the 6th and 7th Battalions were to join them in France.

As one historian has truly stated 'The British army entered the war far less ready than in 1914. Behind the BEF of 1914 lay ten years of steady preparation and thought, while behind the BEF of 1939 lay two years of controversial, confused and scrambling modernization'.[1] Montgomery, then commander of one of the BEF's five regular infantry divisions, was equally condemnatory, 'It must be to our shame that we sent our Army into that most modern war with weapons and equipment which were quite inadequate . . .'[2]

The winter of 1939/40 was that of the 'Phoney War'. The small British Army, deployed in the north with French troops on either side was, as in 1914, tiny in comparison with the eighty French divisions. Strategically, therefore, the French once again called the tune, but this time it was a defensive one. Having over-learned the lessons of 1914-18, French thinking was poisoned by the memory of their losses. Their advocacy of a defensive battle, based upon the steel and concrete of their Maginot Line upon which the Germans would smash themselves, was as fervent as their previous faith in the unremitting offensive. As a break from training and digging, the 1st Battalion spent a short sojourn in that Line, near Metz, where it patrolled and sometimes exchanged small-arms fire with the Germans. There is no record of anyone being hit.[3]

★★★

However, the 1st Battalion was not to fight in France. To everyone's surprise, on 17 April 1940 it crossed the Channel back to England together with the other two battalions of 15 Infantry Brigade, 1st York and Lancasters and 1st King's Own Yorkshire Light Infantry. Five days later these Green Howards were in warships bound for Norway.

This had come about as follows. In order to cut off German supplies of the vital Swedish iron-ore and control the northern sea-route, the Allies had planned to lay mines in the ice-free Norwegian territorial waters; if Germany retaliated by invading the neutral Scandinavian countries, the Allies would occupy the more northerly Norwegian ports. But the Germans struck first. On 9 April they occupied Denmark and began to invade Norway, landing forces at Oslo, Bergen, Trondheim and Narvik, the latter the ice-free outlet for the iron-ore. This confronted the Allies with the task of ejecting them, using forces neither trained, equipped nor properly supported to oppose what might be amphibious landings.

One expedition was mounted against Narvik and two others, each by half-trained Territorial brigades, to envelop and recapture Trondheim. The first of these, landing at Namsos, was quickly crushed as the troops began to move towards Trondheim. Much the same happened to the second brigade, brought ashore on the night of 19/20 April at Aandalsnes, two hundred miles to the south as the crow flies. The task of this brigade had been changed to securing the road and railway junction at Dombaas, some sixty miles inland, and then making contact with the Norwegians, retreating northwards from Oslo up the Gudbrandsdal towards Lillehammer. By the evening of 23 April that brigade's survivors numbered a mere 300.

That same day the first troops of 15 Brigade landed at Aandalsnes.[4] Two days later, on 25 April, after a rough thirty-six hours crossing the North Sea, the cruiser and three destroyers carrying the 1st Battalion were steaming up the still waters of the fiord, the remote serenity of its ice-capped mountains impressing at least one onlooker. Then, around another bend, there appeared the red glare of the still burning Aandalsnes, bombed twenty-four hours before.[5]

The next morning the Green Howards, the last battalion to arrive, were moving south by train to Dombaas, all their equipment having been speedily

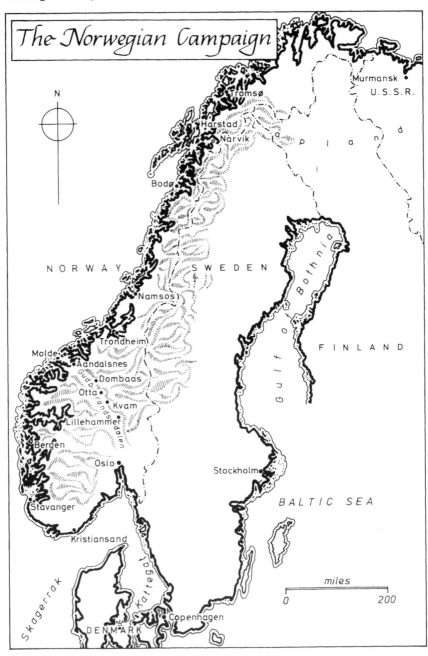

The Norwegian Campaign

manhandled from the ships. The Brigade possessed no transport at all, only half a light anti-aircraft battery, no field artillery and just one improvised company of French 25mm Hotchkiss anti-tank guns, manned by men drawn from the three infantry battalions and commanded by a Green Howard. There were, of course, no tanks; and the Germans had complete command of the air; daylight lasted for eighteen hours.

When the Green Howards reached Dombaas on 26 April the last of the Norwegian troops were withdrawing through the Yorkshire Light Infantry. That battalion had for nearly forty-eight hours been holding off repeated enemy attacks at Kvam, fifty miles to the south, during which the anti-tank company had claimed a couple of tanks, the first so destroyed by the British Army during the war. Some two miles to the rear of the KOYLI were the York and Lancasters, and that night the Light Infantry, exhausted and with many casualties, were pulled back through them. At the same time a Green Howard company and the carrier platoon (less, of course, its carriers) were sent forward in Norwegian trucks to help the York and Lancasters. Meanwhile the rest of the Green Howards were moved by train to prepare another rear position just south of Otta, a town half-way between Kvam and Dombaas.

That Green Howard company with the York and Lancasters had a foretaste of what was to happen later. In the deep valley, never more than a mile wide, the fir-tree covered hillsides lay thick in snow that softened during the day and froze at night to a crust insufficiently hard to support a man's weight. Because digging was difficult, stone sangars were built, deadly when splintered by high-explosives. Although there were a few Norwegian ski-troops operating on the flanks, it was not difficult for the Germans, properly equipped for mountain warfare and showing astonishing dash and initiative, to get above and around the British positions.

Throughout 27 April that company fought with the York and Lancasters. In the end it withdrew in the dark across the snow-covered and broken hillside to take up a covering position to assist the now battered Battalion, an episode that cost the Green Howards its carrier platoon, surprised and slaughtered by German armoured cars.

The intention was that Otta would be held by two battalions, but the York and Lancasters had suffered such heavy losses that the main burden fell on the Green Howards, fighting in a naturally strong defensive position astride the river. Lacking one company that had been moved off to cope with a threat from the left flank, the Battalion throughout 28 April stood alone, covering a wide front against a force of seven battalions of German infantry and one of ski troops, strongly supported by a machine-gun battalion, light tanks and artillery, some 9,000 men in all. And for much of the day the Green Howards were bombed and machine-gunned from the air.

At 2100 hours the Battalion began to extricate itself, a hard task indeed. Despite the confusion inevitable in any withdrawal across broken country in the dark by troops, most of whom were closely engaged with the enemy, few men failed to get back to Dombaas; some travelled by Norwegian truck, some by train and some made the long journey on their feet. Of those that

Left: Germans shelling positions of 'C' Company of the 1st Battalion in Norway in 1940. The origin of this photograph cannot be traced but is certainly German.

did become separated, then and later, several individuals during the course of a series of extraordinary escapades made their way home.

After more fighting around Dombaas on 29 April, the Battalion that night began its final journey back to Aandalsnes where it would be embarked for England. The British Government had admitted that its ill-judged adventure had failed. Some Green Howard companies travelled by road and some by train. For the latter it was to be a foul journey. At 0100 hours the train crashed in a tunnel, setting fire to the wagon carrying the ammunition reserves. The explosions killed and injured many Green Howards, but the survivors, dog-tired and burdened by their heavy weapons, had within forty-five minutes collected themselves and begun the seventeen-mile march along the icy road to the next tunnel, the only cover against the enemy aircraft whose arrival could be expected with the dawn. Happily another train awaited the Green Howards at the second tunnel and there they sheltered until nightfall before continuing their journey to the port.

By the light of burning villages the Battalion embarked in the early hours of 2 May for what was to be an uneventful journey home. A similarly ill-equipped and makeshift expedition, but on a larger scale, was evacuated from Narvik five weeks later. This Norwegian fiasco brought down the Chamberlain government; it was somewhat ironic that the new Prime Minister, Winston Churchill, in his previous capacity as First Lord of the Admiralty, had been responsible for it all.

The contrast between those unfortunate Territorials and the regulars of 15 Brigade had been only too marked. The former were splendid men, but in General David Fraser's thought-provoking words:

Few men are born heroes. Few are incorrigible cowards. Most can be either; and to help them towards the former rather than the latter state an army uses leadership, discipline and training – a mix which produces confidence and pride.

Without these characteristics, he continued, soldiers cannot behave well in battle, and the failure is not their's but the system which has placed them there unprepared.[6]

Right: The Colonel-in-Chief, H. M. Haakon VII, King of Norway, accompanied by the Colonel of the Regiment, the then Lieutenant-General H. E. Franklyn, visits Richmond on 25 August 1943.

Trained and commanded by Lieutenant-Colonel 'Robin' Robinson, later in the war a major-general and afterwards his Regiment's Colonel, those Green Howards who did so well at Otta, many of them reservists with Palestine or Frontier medals, had in plenty that confidence and pride. It was sad that their fine performance was for years forgotten amid the disappointments of that campaign.

★★★

Those Norwegian battles forged yet another link between the Green Howards and Scandinavia, the home of the ancestors of so many North Riding men. Their Queen Alexandra who had, in June 1914, put the seal on her thirty-seven year connection with the Regiment by becoming Colonel-in-Chief, had died in 1925. In 1942 her son-in-law and nephew, King Haakon VII of Norway, who had remained with his government and forces until the last Allied troops left his country, became in turn Colonel-in-Chief. In August 1958 his son, King Olav V, who as Crown Prince had stayed with his father, succeeded him. And then, in early 1992, King Harold V on the death of his father, King Olav, became the fourth member of his family to hold the appointment.

★★★

Just eight days after the 1st Battalion left Norway behind, the long-awaited German offensive overwhelmed the still neutral Belgium and Holland. On 10 May 134 German divisions, ten of them panzers, thrust westwards with paralysing speed, supported by terrifying Stuka dive-bombers. Ahead of them, parachute and glider-borne troops, a new and unexpected weapon, seized supposedly impregnable fortressess and captured undamaged bridges.

It had been assumed that, as in 1914, the main German thrust would cross Belgium to hit the Allied left flank, and plans had, therefore, been laid that the BEF, with the French First Army on its right, would advance when the Germans attacked so as to hold what would be a shortened line running south from Antwerp to Namur, thus linking up with the Belgian Army to the north. Instead, however, the main German onslaught swept through the wooded Ardennes and then swung north-easterly towards the Channel ports.

The result was that, on 16 May, the BEF began its long withdrawal towards the sea. The 50th Division, after covering the retreat of the forward corps, as part of the GHQ reserve, was then directed towards the threatened southern flank. The ensuing fortnight's fighting was confused indeed: Captain Synge, whose meticulous research has uncovered much of what then happened, was still obliged to admit his difficulty in putting together 'a coherent story of those chaotic days'.[7] The 4th and 5th Battalions were almost constantly on the move, often without rations; at times they found themselves out of contact with the headquarters of their 150 Infantry Brigade; companies and platoons were often cut off, fighting on their own.

The potential of those raw and still half-trained units soon became clear when, already exhausted and closely pressed by the enemy, they showed that they could march thirty miles in the day over *pavé* roads crammed with refugees on their way from Belgium towards Arras. Not a single man fell out in either Battalion. Mechanized infantry they may have been, but they could still use their feet. Then, in and around Arras, they were to display the ruggedness in defence that became their hallmark.

At Arras had come together FRANKFORCE, named after its commander, the Green Howards' Colonel, Major-General W.E. Franklyn. His troops were the 5th and 50th Divisions, together with 1 Army Tank Brigade, the only such formation in the BEF. Pivoting on Arras, FRANKFORCE's task was to counter-attack by swinging anti-clockwise around the west and south of the city so as to stem the German drive towards the Channel ports. On 21 May Franklyn launched his attack which at first succeeded brilliantly. The Germans were swept aside, at least twenty of their tanks destroyed and 400 prisoners taken. But FRANKFORCE was hopelessly outnumbered. Against it were deployed two panzer divisions, one of them commanded by a General Rommel. Counter-attacked, FRANKFORCE had to pull back, but the German onrush was delayed.

Neither of the two Green Howard units took part in this counter-attack, but 5th Battalion, helping defend Arras itself, lost men when the city was subsequently heavily bombed. To its west was the 4th Battalion, holding the bridges over the River Scarpe, and there it fought for the next three days. At one time all but encircled by the enemy, the the 4th lost heavily, including almost the whole of a company, but the Germans had heavy losses as well. To those soldiers of 1940, the nearby place names were still familiar. One true Yorkshireman, told beforehand that he would be fighting that night at Vimy, had responded 'Vimy Ridge, Sir! It's just like being asked to play for England at Lord's'.[8]

Also present at this Arras battle were the 6th and 7th Battalions, part of 23rd Division, one of three second-line Territorial formations which had landed in France only the previous month to supply working parties to build airfields. Without any form of artillery or mortars, they averaged only two Brens per company. Many men had hardly fired a shot on the ranges. Hardly of the calibre to cope with *Panzer* attacks, these Green Howards were nevertheless deployed to plug gaps in the line along the Scarpe.

On 23 May FRANKFORCE, all but surrounded by German armour and in danger of being cut off, was ordered to withdraw northwards, a tricky manoeuvre skilfully accomplished. By 27 May, after a number of confused moves, 150 Brigade found itself in Ypres, a city only just restored from the previous war's ruins. There the 4th and 5th Battalions, together with their old friends in the 4th East Yorkshires, were to fight hard for forty-eight hours as part of the now rapidly contracting bridgehead around Dunkirk, the last port left in British hands. Once again, the names of nearly every village recalled the deeds of their fathers: Menin, Zillebeke Lake and Poperinghe. Some old soldiers, who had fought in the Salient more than twenty years before, now found themselves taking cover behind their comrades' graves among those war cemeteries that so densely patterned the Flemish countryside.

But the evacuation by small boats from the Dunkirk beaches had already begun. The remains of the Belgian Army, holding the left flank, had surrendered on 27 May. Struggling through congested roads, the 4th and 5th reached the final close perimeter on 30 May. There they repelled a number of violent German attacks, losing even more men. When they finally trudged through the thick sand of the squalid beach, littered with abandoned equipment and blanket-shrouded bodies, the Green Howards still carried their weapons, holding them above their heads as they waded out to the boats. Not until the evening of 2 June did those two units get away, among the last troops to leave.

On the previous day, the 6th Battalion had embarked, also in a formed and disciplined body. This untrained 'labour' unit had done well. Armed with little more than their rifles, on 23 May they had been directed to Gravelines on the coast. There, with the help of a few cruiser tanks, they had somehow held out against 1st Panzer Division for two days, so 'considerably delaying' the enemy attack from the west, as the citation to Lieutenant-Colonel W.R. Steel's subsequent bar to his DSO was to testify.[9] Theirs was a magnificent achievement that lifted morale and helped lay the groundwork for a superb unit. As in the First World War, much more was to be heard of that 6th Battalion.

But it had been a calamitous campaign. Never before had the British Army known such an overwhelming defeat. The Germans now occupied or controlled all western Europe. Britain, completely alone, awaited invasion. The bulk of her army had been brought off the Dunkirk beaches, but without guns, tanks, vehicles and every other type of equipment.

★★★

During the next eighteen months no Green Howard unit was to be in contact with the enemy. At first, those five battalions, back from Norway and Dunkirk, stood almost weaponless around the British coasts awaiting an invasion which the Battle of Britain, fought in the summer and autumn of 1940 by the Royal Air Force and Anti-Aircraft Command, made impossible. Soon, however, equipment began to arrive and time could be spared for proper training.

In the reorganization that followed Dunkirk, 23rd Division was dissolved. As a result, 69 Brigade, with its 6th and 7th Green Howards and 5th East Yorkshires, joined 50th Division, a welcome addition indeed. Such was the need for manpower in the many more specialist units required in this Second World War, that no Green Howard unit, other than the six pre-war regular and Territorial units, was to see action. Several were raised, but later either disbanded or used as second-line garrison units. Only the 10th Battalion, formed in June 1940, was more fortunate. Such was its high standard that, in May 1943, it was selected for conversion to a parachute battalion. Every officer but one, together with 300 men, at once volunteered, and in due course it became the 12th (Yorkshire) Battalion the Parachute Regiment, a unit that in due course won high distinction; sadly, however, it lost all direct connection with its parent regiment.

To help repel the invasion that never came, another large body of Green Howards was to be raised. The Home Guard, brought into being in May 1940, in the end included thirteen Green Howard Battalions, a total of 15,000 officers and men, most of them staunch members of the Regiment, proud indeed to wear its cap badge.

***

Hitler's victory in Europe was the signal for Mussolini to snatch his share of the spoils. In September 1940 a strong Italian Army began to edge its way eastwards from Libya into Egypt, thus threatening Britain's Middle East oil supplies. The response was dramatic. In December 30,000 British Empire troops sent an Italian army five times its size reeling back 500 miles with the loss of 130,000 prisoners. It was the boost to morale so badly needed both by the nation and its army.

This victory was largely due to Churchill's making one of the boldest decisions of his valiant career. Disregarding the danger of imminent invasion, in August 1940 he despatched to the Middle East some of the very few armoured regiments available at home. Then, in April 1941, 150 Brigade followed the armour, as did the rest of 50th Division a few weeks later.

The survivors of those Green Howards never forgot the magnificent hospitality they received from the South Africans as their convoys rounded the Cape of Good Hope. But, having arrived in the Middle East, it was to be many months before they saw proper action, although the war in the Mediterranean had begun to go awry even before 50th Division left the UK. In February 1941 Rommel's newly arrived *Afrikakorps* had swept the British back to the Egyptian frontier; then German divisions overran both Greece and Crete, crushing in the process the outnumbered British troops sent from the Middle East to help save them.

To cope with these widespread crises, Green Howards saw much of the Middle East: Cyprus, Egypt, Syria and Palestine were all visited at different times; the 6th and 7th Battalions motored as far as the Kirkuk oilfields in Iraq to counter the danger of a German southwards thrust through Iran, Hitler having made that near inconceivable error of invading Russia in June 1941.

Above: General Sir Edward Bulfin. Oil painting by John St. Helier Lander.

At last, however, in February 1942, 50th Division entered the line south of Gazala, making ready for the next Axis onslaught, Bengazhi having been won and lost yet again in the see-saw of that rapidly moving desert campaign, fighting in which 4th and 5th Battalions, detached from their Division with 150 Brigade, had for some weeks already been experiencing their first taste of desert warfare. True professionals now, all four battalions had benefited from the strong leavening of keen young regular officers and senior NCOs who had joined just before the battalions left England.[10] It was fortunate that as yet losses among regulars had been slight.

During the following months the Green Howards learned how to fight and live in the desert. They dug, wired and laid mines, and patrolled aggressively against the Germans and Italians. Sometimes they existed on as little as half a gallon of water daily, profiting from the water discipline that had been drilled into them during their training. And despite the thirst, the flies, the choking dust, the terrible food, the cold of winter and the grilling *khamsin* wind of the summer, the desert had a peculiar attraction for many of those who fought there, even amid the beastliness of war.

On the night of 26 May Rommel's panzers struck once again, sweeping around the south of the Free French at Bir Hacheim and hitting the British rear – the famous Knightsbridge area. In doing so they bypassed the so-called 'box' held by 150 Brigade Group, ten miles distant from the French and six from 69 Brigade to the north. At first the Brigade was left in comparative peace, but gradually, as Rommel realized that success depended on opening the supply routes which it covered, attacks against the box's twenty-mile perimeter intensified. For five days the 4th and 5th Green Howards, together with the 4th East Yorkshires and their associated gunners, sappers and tankers, doggedly defended that piece of bare desert – later to be known as 'the Cauldron' – against six separate attacks from Rommel's panzers and dive-bombers. Then on 1 June tanks, self-propelled guns and infantry were unleashed from every direction against 150 Brigade. In turn platoons were overrun, their ammunition finished. That morning their Brigadier, Bill Haydon, who had led them in France, was killed: when the battle began his words had been 'here we bloody are and here we bloody stay, nolens bloody volens'.[11] And so those Territorials did.[12] But by midday it was all over and the German supply columns were able to pass through. 150 Brigade had ceased to exist. Rommel himself was there to see its end.

> Yard by yard the German-Italian units fought their way forward against the toughest British resistance imaginable. The defence was conducted with considerable skill and, as usual, the British fought to the last round of ammunition.[13]

With these words Rommel paid tribute to 150 Brigade.

Meanwhile, behind 150 Brigade a massive tank battle had been raging,

Left: C. S. M. Stanley Hollis painted by Albert Lembert of Darlington. His was the only Victoria Cross to be won on 'D' Day.

one that was to continue for a further two weeks until the British armour was all but destroyed. By then the 6th and 7th Battalions had already been fighting for two exhausting weeks both in and outside 69 Brigade's box. As yet, however, casualties had been light except on 5 June when the 7th Battalion lost heavily in supporting a massed and unsucessful tank attack which failed to retake the old 150 Brigade position.

With the Germans and Italians now all around them, the withdrawal of what remained of 50th Division became inevitable, an operation that was to be conducted in a masterly manner. Instead of trying to break out eastwards, with superb boldness units struck forward in the dark through the Italians and their minefields. Small columns of vehicles then swung south to make a wide detour around Bir Hacheim, now also in enemy hands, towards the Egyptian frontier. Although much kit was lost, the sound training and discipline and the magnificent *esprit de corps* of those North Country Territorials had kept casualties down.

Having sorted themselves out near the frontier, the 6th and 7th Battalions helped cover the withdrawal of the shattered Eighth Army as it fell back even farther to Mersa Matruh. In charge first of the 6th Battalion and then, from 26 June, of 69 Brigade was a famous Green Howard, Brigadier 'Red Ted' Cooke-Collis, one of the finest fighting commanders produced by the war, his nickname due not to his complexion but his conspicuous devotion to his rarely discarded red-banded service dress cap.

Outside Mersa Matruh on 27 June 69 Brigade, in an attempt to delay the German advance, counter-attacked but hit a vastly superior enemy force. Surrounded once again, the 7th Battalion was all but annihilated, but the 6th, its strength now about halved, managed to escape and fight a superb series of delaying actions back to El Alamein, just short of Alexandria.

Brought up to strength by a few survivors of the 7th Battalion and a mass of reinforcements, some of whom arrived in the middle of the battle, the 6th was to toil at laying minefields and digging defences in the desert summer heat from the time it reached Alamein. There was no time for training and, as was to be the case so often in the future, company and platoon commanders at times hardly knew the names of the men they led into action.

Then, at Alamein during July, with its right secured by the sea and its left by the Qattara Depression, thirty miles to the south and impassable to tanks, Eighth Army halted the panzers, a battle in which 6th Battalion played a superb part. A German victory would have opened the door to Egypt and brought about the loss of the Middle East, but already exhausted units such as the 6th Green Howards and the 5th East Yorkshires in some way did the impossible at what was to be known as the First Battle of Alamein. In command for just a few days before he was wounded was George Eden, later, after his retirement as a brigadier, to succeed Major-General Robinson as Colonel of the Regiment.

But at Alamein at the end of July 1942, the British were at the nadir of their fortunes, and the Green Howards in two years of war had still to fight a winning battle. It says much for the groundwork laid in front of Gazala that those two Green Howard units fought on as they did for the next two and a half years.

★★★

With hindsight it can be clearly seen that the arrival in Egypt in mid-August of that trenchant little general, last encountered addressing 1st Battalion in Palestine, was to prove the turning-point of the war. Soon General Montgomery was known by reputation to all and by sight to most of his Eighth Army. One of his techniques was to gather complete brigades around him and tell them in plain language just what he expected of them. North Country men rarely respond to pep-talks from strange generals, but Monty had a different way of saying these things. One officer remembered their reaction: their shoulders went back and they seemed to grow a couple of inches. Another lesson Green Howards learned was that Monty would take care of their lives. Visiting the 6th Battalion, then temporarily under command of 7th Armoured Division, and finding them occupying a defensive position with neither anti-tank guns nor other supporting weapons, he ordered the Battalion back to Alexandria with the words 'I will not have any men put into action without proper equipment'.[14] The 6th Battalion was to find its needed equipment: Montgomery's arrival had coincided with that of shiploads of modern *matériel*.

Also in the Delta was the 7th Battalion, absorbing vast drafts but carrying out little training because of the demands for working parties. Nor was the 6th given time to organize itself; no sooner had it been pulled back than two of its three rifle companies were returned to the desert to reinforce other depleted units of 50th Division. This treatment, for which soldiers have a terse expression, gave the division no time to recover from the damage it had so recently suffered. Because of this it was not in good shape when Montgomery's decisive offensive at El Alamein began. Other and freshly arrived divisions, on the other hand, had been given ample opportunity to rehearse their roles.

Across the thirty-mile front line the opposing infantry divisions faced one another from strongly fortified and mined positions; behind the infantry waited the armour. Montgomery's plan was make a breach in the enemy defences in the north through which he would then pass his armour. At the same time a subsidiary and southerly thrust, of which 50th Division formed a part, would attempt to attract Rommel's panzers away from the main offensive.

Throughout 24 October the Green Howards, who had already been in the line for the past two weeks, had listened to the unending roar of the battle raging to their north. Having learned that the northern attack was gaining ground, their turn came the following day. On 25 October 69 Brigade went into the attack, its two forward battalions astride the mile wide Munassib Depression, the 5th East Yorkshires on the right and the 6th Green Howards on the left. In reserve was the 7th Battalion.

El Alamein, in its opening stages, was very much a First World War battle. Infantry on their feet, covered by massed guns and mortars and sometimes tanks, clawed their way to their objectives through wire and machine-guns, hammered as they did so by the enemy's bombing and shelling. But it was not attrition pure and simple as it had been in 1917 and 1918: the tank was now fast moving as well as reasonably reliable, a weapon able to exploit any gaps the enemy might make.

So it was that the 6th Battalion's attack much resembled those made by their fathers a quarter of a century before, their losses on much the same scale. Thirteen officers and half or more of the rank and file in the three rifle companies were to be either killed or wounded; as before, bayonets, bombs and even pistols were used among the enemy dugouts and slit trenches. Only when the last Green Howard company was down to a third of its original strength was the attack halted, its primary task completed.

On 3 November the breakthrough in the north occured. All along the line, as the Italians and Germans began to withdraw, Eighth Army started on its 1,500-mile pursuit westwards towards Tunisia. During the first four days mobile columns based upon the three battalions of 69 Brigade took part in the initial stages of this exploitation, advancing until they were level with Mersa Matruh and rounding up many thousands of weary prisoners. Then, with the remainder of their division, they were granted a badly needed respite so as to fashion themselves again into the powerful fighting machine they once had been.

The Green Howards' part at Alamein had been a minor one, but it had cost them dear. They had done just what had been asked of them and done it in fine style.

★★★

By February they were again ready, their task this time to breach the Mareth Line. It was to be yet another old-fashioned slogging match.

A miniature Maginot Line, these Mareth defences had been built by the French in pre-war days to protect their Tunisian frontier when Mussolini was planning further to expand his Italian Empire. Based on the deep and wide Wadi Zigzaou, the Mareth Line was a complicated system of concrete pillboxes, cleverly constructed trench systems and deep tank ditches backed by wide belts of wire and mines and manned by both German and Italian infantry.

Montgomery's plan for dealing with the Mareth Line was characteristically simple. The 50th Division would assault it on a narrow front, while the New Zealand Division, strongly supported by armour, would swing around the left in a subsidiary attack.

With what an onlooker described as 'clockwork precision',[15] the three units of 69 Brigade first drove in the enemy outposts to the Line during two nights of bitter and costly hand-to-hand fighting. Then, at 2200 hours on 20 March, under cover of the full weight of the corps artillery, the 7th Green Howards moved forward to open the Battle of Mareth proper. Its objective was the 'Bastion', an outlying strongpoint that enfiladed the ground over which the three Durham Light Infantry battalions of 151 Brigade were to make the main attack.

Above: Soldiers on exercise carried on a Valentine tank on 12 March 1943.

Right: A lone soldier guards a batch of German prisoners captured at Mareth on 29 March 1943.

Right: Brigadier 'Red Ted' Cooke-Collis, in command of 69 Brigade, watching his troops move forward in the Gabes Gap area.

A hail of artillery, mortar and machine-gun fire met the 7th Battalion as it moved into the assault. Realizing the seriousness of the situation, the commanding officer, Lieutenant-Colonel 'Bunny' Seagrim, pushed to the front and took personal charge of a team attempting to place a scaling ladder over the twelve-foot-wide anti-tank ditch. The first man across, he attacked two machine-gun posts with his pistol and grenades, killing or wounding ten or a dozen Germans. It was largely due to Seagrim's inspired courage that the 7th Battalion captured the 'Bastion' and then held it throughout the following day against a succession of counter-attacks.

In the meantime the three Durham battalions had seized a shallow bridgehead over the Wadi Zigzaou, but not enough tanks and anti-tank guns could be brought across it to halt the strong German armoured counter-attacks. Into the bitter 'dog fight' that ensued, the 6th Battalion was brought forward from reserve during the evening of 22 March. Lying under heavy shellfire and surrounded by the apparent disorder inseparable from such occasions, one confident private soldier was heard to encourage his mates with the words, 'Don't you worry, lads. Monty and Red Ted have got this show laid on, and they won't move us from here until everything is properly organized'.[16]

In the end Montgomery was obliged to halt 50th Division and reinforce the New Zealand left hook. The Mareth Line was breached but a large part of the enemy got away, to make a further stand sixty miles back on the next defensive line, the Wadi Akarit. On 6 April, flanked by 4th Indian and 51st Highland Divisions, 69 Brigade helped win the final major battle in the North African campaign. Described by Montgomery as 'the heaviest and most savage fighting we have had since I have commanded the Eighth Army',[17] the Green Howards again suffered heavily but again did equal damage to the enemy. With the East Yorkshires the two battalions either killed or took prisoner practically an entire Italian division.

The next day, from their high ground, they watched the British armour sweep northwards. A month later 250,000 Germans and Italians were to surrender. Between them Eighth and First Armies (the latter had landed in Algeria in October) had cleared the enemy from North Africa. It was the first major victory for the Western powers, one in which the Green Howards had taken their full share.

Mareth and Wadi Akarit had cost the 6th and 7th Battalions 500 officers and men killed and wounded, among them 'Bunny' Seagrim, hit mortally as again he led his 7th Battalion from the front at Wadi Akarit. Thirty gallantry awards were won by the two battalions for these two battles,[18] one of them the Victoria Cross by Lieutenant-Colonel D.A. Seagrim for his superb bravery and leadership at Mareth. He died before the news could reach him. The Seagrims were a gallant but ill-fated family: his elder brother, an officer in the Indian Army, was to be awarded a posthumous George Cross for his work behind the lines in Burma.

After advancing into Tunisia as far as Enfidaville, 50th Division was withdrawn all the way back to the Canal Zone in Egypt to make ready for its next step towards helping rescue Europe from Hitler's tyranny.

10. SECOND WORLD WAR: SICILY TO THE ELBE, 1943-5

Before 1939 the recruiting slogan had been 'Join the Army and See the World'. Those many regulars still remaining in the 1st Battalion when it next left England were to see quite a lot more of it. For three years their life was to be one of near incessant travel.

After its hasty exit from Norway the Battalion had spent the following two years guarding the shores of the British Isles from invasion and undergoing the effective training, based on the lessons learned during the Dunkirk campaign, which the Army was now undergoing. Scotland, England and Northern Ireland in turn saw the 1st Battalion. In Ulster, where their Colonel, now Lieutenant-General Sir Harold Franklyn, commanded, the principal reason for their presence was the fear of a German incursion into the the Republic of Ireland.

The Battalion's eventual departure from the British Isles as part of 5th Division in March 1942 was brought about by Japanese aggression in the Far East. The attack on the US fleet at Pearl Harbor in December 1941 had brought the USA into the war but had also spelt peril for Asia. Malaya, the British naval base at Singapore, Hong Kong, the American-occupied Philippines, and the then French Indo-China and Dutch East Indies had all in turn fallen to the Japanese, whose advance through Burma soon threatened India itself.

Landing at Bombay in May, the 1st Battalion travelled across the sub-continent to Ranchi. That summer, however, an even greater danger arose and 5th Division was switched back westwards, moving from the sweltering Indian plains to the freezing Iranian uplands to help defend the oilfields from yet another and more serious German advance from the Caucasus. When that threat also receded, the Battalion motored south by way of Baghdad and Palestine to Egypt where it arrived in March 1943.

★★★

During its short stay in India the 1st Battalion had, for only the third time in its history, met the 2nd Battalion. Still on a peace footing both in theory and in practice, its fate was to all but mirror that of the 1st Battalion during the 1914-18 War. One of the few Regular units to be kept in India for internal security duties, the Regular element of that unfortunate 2nd Battalion quickly disappeared to help man newly raised units in India and elsewhere. Most of those who remained sought, often with increasing desperation, for a more active part in the war. (The author writes with personal knowledge.) Their places were frequently taken by somewhat untrained men from home, many far from pleased at finding themselves in India. Nevertheless, the Battalion did well when sent to Razmak at a week's notice in October 1942. There the excitements, discomforts and tactics of Frontier warfare had in no way changed. A few men were hit by tribesmen's bullets; for fourteen months the Battalion had a worthwhile task to perform.

At last, in September 1944, the 2nd Battalion was to see something of the war proper. Joining 26th (Indian) Division in the Arakan, south of Chittagong, just before Fourteenth Army took the offensive against the Japanese, it quickly acquired the skills of patrolling in tropical jungle against hardy and skilful enemies. Its strength no more than half that of neighbouring units, it then took part in the almost unopposed amphibious landing on Ramree Island on 21 January 1945, the object of which was to seize airstrips from which General Slim's main attack towards Mandalay could be supported. Crossing from Ramree to the mainland, for the next three months the 2nd Battalion lost men both to the Japanese and to disease, but without fighting a major action. The Burmese campaign brought the Green Howards much discomfort, some casualties but little glory during their service in what was to become known as the 'Forgotten Army'.

★★★

We have jumped ahead however. Earlier in our narrative we left three Battalions of the Regiment, the 1st, 6th and 7th, converging on the Canal Zone in Egypt. There, on the Bitter Lakes, they were in turn to assimilate the skills of amphibious warfare, the art of making an opposed landing from the sea, techniques almost completely neglected during the inter-war years.

Sicily was now the target. The fortunes of war had at last changed. For a year Britain had faced Hitler alone. Now troops of the US Army were to take that first hard step, alongside those of the British Empire, towards helping clear the Germans out of the Europe they had overrun.

The two formations to which the Green Howards belonged, the 5th and 50th Divisions, landed alongside each other before daylight on 10 July 1943, at the south-east corner of Sicily, the task of their XIII Corps to advance along the coast and capture the port of Augusta. No Green Howard unit was among the first waves of troops who waded ashore from their assault craft, and at first the Italian garrison troops offered only slight opposition, but this very soon stiffened. The Green Howards, moved forward into the battle, came up against what were, by comparison, slightly less irresolute Italian field divisions. But with them were formidable *Panzer* formations, including the redoubtable Hermann Goering Division.

After the deserts of the Middle East, Sicily at first seemed unimaginably lovely. Everywhere the towering pinnacle of Mount Etna dominated the attractive pink, white and blue houses clustering among orange and olive groves. Chickens, eggs, fruit and wine could be found to lend variety to the 'Compo' rations. Male relations of friendly and gaily dressed girls hailed the astounded soldiers as liberators. But the countryside was hostile to soldiers who had learned their trade in the desert. Rugged hills, gorges, winding roads easily mined and stone walls, similar to those remembered from the North Riding, posed altogether unfamiliar tactical problems.

In the torrid Sicilian summer, the three dusty and thirsty Green Howard battalions pushed northwards towards Catania from one ridge to the next, capturing droves of Italians but meeting ever more bitter resistance from the Germans. Typical of the fighting was 7th Battalion's attack on Mount Pancali, just short of the Catanian Plain, four days after the landing. The hill was held by a company of Germans supported by machine-guns and a troop of self-propelled artillery. An observer who watched the attack wrote:

> The show put in the shade any demonstration which an Infantry School could produce. The planning was masterful, and the execution, under intense fire, was brilliant. When it is considered that thirty German machine-guns under excellent cover, and cunningly concealed, swept the entire area continuously, and that the climbing of the hill would have been unpleasant under peace conditions, some idea of the stupendous task that was undertaken can be imagined.

A German desert veteran told his captors, 'I have been against many British attacks, but that was the finest I have ever seen. I congratulate you!'[1]

As soon as Mount Pancali was firmly consolidated, the reserve company and carrier platoon passed through to continue the advance. Casualties continued to mount. In that sorely tried 7th Battalion, which had already suffered so severely in North Africa, yet another three officers died that afternoon, one by shelling, one in an ambush and one by dive-bombers. This very professional attack opened the way for the Durham Light Infantry of 151 Brigade to pass through the 7th Battalion to relieve the remnant of 1 Parachute Brigade which had dropped the previous night to capture the Primasole Bridge over the River Simeto. In this, the largest British parachute operation as yet mounted, most of the aircraft went astray and no more than a handful of airborne soldiers reached their objective. It took the DLI two days of hard fighting to reach the survivors.

The way forward towards and beyond Catania should now have been clear, but German paratroops, tough veterans of Crete and Russia, were being flown in to Catania airfield. On the night of 17 July these formidable opponents, some occupying the concrete and mined airfield defences, halted an attack by one of 50th Division's brigades. Directed around the German left flank in the dark across unreconnoitred ground, the 7th Battalion then met heavy machine-gun fire and finished the night with only 115 men surviving in its four rifle companies.

Two days later, units of both 5th and 50th Divisions made further attempts to get forward. Attacking almost alongside each other, the 6th Battalion made some progress, but the 1st Battalion was hit by withering machine-gun and mortar fire in a night attack laid on so hurriedly by its brigade that company commanders never saw the ground beforehand and issued their orders as they advanced towards the start-line. One still remembers the occasion as a classic example of 'order, counter-order and disorder'. When dawn broke the Battalion found itself overlooked at short range and pinned down. Not until the next night could the survivors, many of them wounded, be brought back.

Among the wounded the 1st Battalion was forced to leave in enemy hands was one of the company commanders, Captain Hedley Verity, perhaps England's greatest slow left-arm bowler and a very gallant and popular Green Howard. He died in a German Army hospital. Captain Laurie Hesmond-halgh, a Cambridge boxing 'blue' and a fine cricketer, was also mortally wounded there. Before the war, the commanding officer, Lieutenant-Colonel Arnold Shaw, a madly enthusiastic cricketer himself, had known Verity well. Not a man to neglect any chance of strengthening his team, among the other distinguished cricketers he had gathered into the 1st Battalion side was the future Yorkshire and England captain, Lieutenant Norman Yardley, another fighting soldier, later to be wounded in Italy. Wherever the Green Howards moved during that world tour, they took with them the cricket mat and a few bags of cement.[2]

★★★

The failure to make progress against the formidable defences of the Catanian Plain caused Montgomery to switch the main force of his offensive to the west of Mount Etna. For the next fortnight the weary 6th and 7th Battalions helped hold the Primasole bridgehead, losing men from shelling and mortaring. Meanwhile the 1st Battalion was granted two short respites during one of which it absorbed a draft of 100 men from the Royal Sussex Regiment (the reinforcement of units in this way was becoming ever more commonplace). On 4 August the Battalion crossed the River Simeto some miles to the west of Primasole. Advancing twenty-one miles in the first twenty-four hours, it mounted a full-scale but unopposed night attack and fought two sharp engagements against an at last retreating enemy. And so during the next week the 1st Battalion pushed on through hills even more ruggedly difficult than those they had clambered over south of Catania, the steady drain of losses continuing, some exacted by the new and heavily armoured German Tiger tank holding the roads and tracks, its armament the formidable 8.8cm gun. At the same time, the move by Eighth Army's other corps around Mount Etna had also gone well and now threatened the German rear.

When the 1st Battalion moved forward, so did 6th and 7th Battalions towards and beyond Catania, fighting over country similar to that through which their regular unit was toiling. By now malaria, Sicily's scourge, was also taking its toll, with men going sick in ever-increasing numbers.

But the end was near. On 17 August the 6th and 7th Battalions heard that General Patton's Americans, sweeping around the north of the island, had entered Messina. By then the 1st Battalion, having lost Arnold Shaw on his promotion to brigadier, had already been withdrawn to prepare for the next step in the war, the landing in mainland Italy itself. At the same time the 6th and 7th Battalions received news that was hard to believe: their division was ordered home to take part in the invasion of north-west Europe. It was easy to understand why it had been chosen. As *The Times* weekly edition wrote:

. . . the sheet anchor of the army was the veteran 50th Division. They had the hard, dirty work . . . They got less public mention than some other divisions because theirs was the unspectacular flank of the front. They plugged on, learning the new warfare the hard way . . . Tyne and Tees may well be proud of them, for they are a grand division.[3]

It had been an unpleasant six weeks. The scale of the Regiment's losses may be judged by the fact that among the killed and wounded in the 7th Battalion were the second in command, all the company commanders, the carrier and several other platoon commanders, the intelligence and medical officers, and the RSM.

★★★

The 1st Battalion had just eighteen days to refit and absorb reinforcements before it tackled its next challenge, the invasion of the mainland of Italy. As part of the reserve brigade of 5th Division, its two-hour crossing of the Straits of Messina on 3 September met no opposition; some described it as a 'pleasant little cruise'. Touching down near Reggio on the 'toe of Italy' and following a sign that read 'All traffic for London straight on 1,870 miles',[4] the Green Howards began that long journey up Italy's rugged spine. This southerly landing by Montgomery's Eighth Army was a subsidiary thrust to draw attention away from the main Allied attack, the hard-fought landing by the largely American Fifth Army six days later at Salerno, 200 miles to the north.

At the start life was not too bad. The weather was good and the enemy were fighting no more than a delaying action. The Battalion's first skirmish, two days after landing, in which fifty-five Italians were captured, was the last by Green Howards against Italians. Mussolini had already been deposed by a *coup* in July, following which the Italians surrendered unconditionally on 8 September, the day before the Salerno landing. The first of the Axis powers had been eliminated.

Moving at times in trucks, more often on their feet and once ferried along the coast by landing craft, the Battalion was often delayed by the ubiquitous mines, booby-traps and demolitions skilfully engineered by the Germans. Not until October, when Eighth Army had been switched across the Appenines to the Adriatic coast, did the Green Howards see any serious fighting. Winter abruptly intervened at the same time. Dust gave way to mud; driving rain swelled the streams and rivers into torrents. It was hardly the Italy of the travel posters. Carving their way forward across the jagged spurs that ran at right angles to the coast, the advancing troops discovered that those picturesque mountain-top hamlets had often been converted to small fortresses. Each river line in turn had to be forced against a formidable enemy.

The Battalion spent one especially unpleasant November fortnight holding a five-mile stretch of the front line overlooking the River Sangro. Together with the divisional artillery, its task was to distract attention from Eighth Army's major river-crossing operation downstream. Freezing and torrential rain flooded what forward trenches could be scraped in the rocky

*Above: The 1st Battalion advancing through the Italian mountains
in the Lanciano sector during the winter of 1943/4.*

soil; to reach them could be a seven-hours' return journey for the Cypriot-led mules now used to get supplies forward. Soon the overworked animals, often burdened by wounded and exposure cases on their return journey, collapsed; either local donkeys, little larger than Alsatian dogs, or the back muscles of the anti-tank platoon replaced them. Battle dress and greatcoats were no protection in such conditions (it was to be another forty years before the army was issued with proper waterproof clothing). Shelling and mortaring rarely eased; the drain of casualties continued.

As always, the Green Howards did what they could. Well concealed and camouflaged tents were erected on reverse slopes. When conditions allowed, men were brought back there from the forward positions for hot food prepared by the company cooks, and even for a rest and dry out. Each evening the same American sergeant appeared with a contact patrol. After a few days one of the company commanders commented upon the fact. 'Gee, major,' came the reply, 'I wouldn't miss this assignment. The only time we ever get any hot chow is when we come over to see your boys.' Once away from their road transport the Americans subsisted exclusively on their unappetising 'K' rations.

The Battalion's next long spell in the line covered the Christmas of 1943, celebrated with cold bully and biscuits. Soon snow lay three or four feet deep. Communications became a major worry. Out almost continuously repairing lines cut by the shelling, signallers lived an especially hazardous life: four were to be killed and two wounded. At one stage when the Germans looked likely to bring up tanks, the Battalion's 17-pounder anti-tank guns (which had replaced the 6-pounders) had to be dismantled and manhandled forward, each barrel a ten-man load. As ever, the infantryman's lot was not an easy one.

By Christmas the Germans had halted the Allies short of Rome along their *Winterstellung* or Winter Line. Fifth Army had been held up along the River Garigliano and Eighth Army just forward of the Sangro, its strength spent in crossing that formidable obstacle. Ammunition was also running short. Because of this Eighth Army went on to the defensive in order to strengthen Fifth Army's next thrust towards Rome. Among its reinforcements was 5th Division, moved across to the west coast in the utmost secrecy.

As the 1st Battalion set off on 6 January, rumour was rife. Incorrigibly optimistic as ever, all were sure that they were bound for home and that the *Queen Mary* was waiting for them in the Bay of Naples. General Montgomery himself had inadvertently provided credibility for the gossip. Ordered back to the United Kingdom just after Christmas to help prepare the coming invasion of north-west Europe, his farewell letter to 5th Division had contained the all too easily misinterpreted phrase 'I hope we shall all meet again soon.'[5] It was not to be. Bloody fighting awaited the Green Howards.

The division's first task was to assist in the assault crossing of the Garigliano on the night of 17/18 January, an operation of immense complexity. On the right, opposite Cassino, Fifth Army's attack failed, but nearer the coast the two leading brigades of 5th Division won a small footing across the river. Only superb work by the gunners, who put down what was described as a 'solid steel wall' of defensive fire,[6] halted the savage German counter-attacks and saved the forward units from destruction. The next night 15 Brigade was brought up to reinforce and extend the attenuated bridgehead. The moment was, according, to the German commander, Field Marshal Kesselring, the greatest crisis faced by the Germans during the Italian campaign.[7]

Crossing by a pontoon bridge, the 1st Battalion had as its objective the village of Minturno, four miles from the river. In a succession of bold but ably mounted company attacks, the village was taken after very heavy fighting. The next day, 21 January, a similar task again faced the Battalion. Advancing in broad daylight under a heavy artillery barrage, the two leading companies first dropped into the valley on the far side of Minturno and then fought their way up the steep slope opposite, to take the small town of Trimonsuoli. Only grim determination won the day, but any elation at success was quickly dampened by news of the failure of the units on either flank: the Green Howards were left holding a dangerous salient exposed to continual shell and mortar fire. Relieved by a battalion of Coldstream Guards, they were that same day thrust back into the now confused battle, its savagery marked by seven more Military Medals won during the fighting on 22 January alone.

Much of the 1st Battalion's success during these savage battles along the Garigliano was due to the high quality of its leaders. Among the several battle-hardened Green Howard regular officers still present were its colonel and all four of its rifle company commanders, the latter men in their early twenties; a number of regular warrant officers and senior NCOs were also still serving. This unusual leaven of experience time and again managed to weld together the hotch-potch of reinforcements filling the Battalion's depleted ranks.

★★★

During the remainder of January and throughout February the 1st Battalion, with two short breaks, helped hold the line, which by then had stabilized around Minturno, activity being confined mainly to aggressive patrolling. During this period General Mark Clark, the American Fifth Army Commander, arrived one day. Among his other remarks was a personal guarantee that no German killed by the Battalion would be replaced as all available enemy reinforcements were being sent to deal with Anzio. A few days later, on 2 March, far from welcome news arrived: 5th Division itself was to move into Anzio. Enough publicity about its horrors had already reached the troops without Clark's adding to it.

Four days after the opening of the Garigliano battle, a US corps, made up of one American and one British division, had landed at Anzio, sixty miles behind the Germans, its task, in conjunction with a southern offensive, to help capture Rome. But over-cautious progress had resulted in stalemate and Kesselring's armoured reserves were able to pin the Allies into a small bridgehead where they withstood a succession of hammer-blows. Then, as further formations were poured in to reinforce failure, 160,000 Allied troops (together with a few nurses) were eventually to battle in that small crammed area, in conditions not unlike those endured on the Western Front in the winter of 1917.

Brought ashore on 6 March, that night already weary Green Howards relieved even wearier Light Infantrymen in 'The Fortress', on the left flank of the bridgehead behind the River Moletta.[8] This 'Fortress' was a complex warren of winding vertically-sided wadis, covered with thick undergrowth and scattered thickly with the foul, stinking debris of war; in places men were entrenched no more than fifty yards away from the Germans and within earshot of their morning roll-call. Most positions were overlooked by the enemy. On one occasion hot goulash was delivered in error by a German ration party to a Green Howard platoon. To stand upright was suicidal. To

Left: Private Caulfield of the 1st Battalion lunches on a salmon sandwich outside his dugout at Anzio. The photograph was probably taken in the rear area.

Right: Men of the 1st Battalion at Anzio prior to the attack on the Moletta River, 22 May 1944.

move off a defined route was to risk death or mutilation in uncharted minefields. Shelling and mortaring by both sides were constant; in one two-week period the Green Howard mortar crews fired 4,000 rounds. Until winter gave way to mosquito-ridden spring the weather was atrocious. Always soaked to the skin and standing in water-logged weapon pits, men's sole relief was a large tot of rum and a pair of dry socks brought up nightly with the rations.

There, in and near the 'Fortress', the Battalion suffered for eleven weeks, broken by short rest periods in the transport lines but still within range of the German guns. Not all the shells contained high-explosives. On 3 April the Green Howards were showered with a variety of propaganda leaflets: one with the caption 'While you die . . .' depicted an American GI fixing his tie as he eyed a half-dressed woman lying on a bed. Discussion centred on whether it was pre- or post-intercourse![9]

On 11 May twenty-five heterogeneous divisions – seven US, five British, four Free French, three Indian, two Canadian, two Polish, one South African and one New Zealand – struck north towards Rome, in fighting as bitter as anything as yet experienced in the west. One week later the Gustav Line was finally broken and Cassino at last fell.

The start of the German retreat was the signal for the Anzio force to break out and cut it off. To the Green Howards fell the unenviable task of attacking the German 4th Parachute Division on the other side of the River Moletta to prevent it interfering with the main American assault scheduled to begin two hours later. The news of this, given to the assembled war-weary unit by an unidentified senior commander, was received with rather less than enthusiasm.[10]

The task of the 1st Battalion, reduced in strength to three rifle companies, was to cross the river near its source and advance across the dunes to the German-held farm, 'l'Americano', 1,000 yards beyond. During the

previous night much of the artillery within the bridgehead had rained shells upon the enemy facing the Green Howards. Then, before dawn, a breaching party crept forward to cut the wire, lay directional tapes and place scaling ladders against the high river banks. Because of heavy casualties from an uncharted minefield, the job was only partially done, but 'D' Company, moving forward at 0445 hours with splendid dash, made good a small bridgehead on the far side of the river. Through it 'C' Company was then passed to capture 'l'Americano', but in storming across the bullet-swept ground, more than two-thirds of its numbers fell. With 'D' Company equally badly hit, 'B' Company was committed piecemeal to help fight off the succession of counter-attacks which continued all day and throughout the next night. When, after a fighting withdrawal, the 1st Battalion was allowed to pull out the next morning, it had lost six officers and 149 other ranks, mostly from the three already understrength rifle companies.

Just three days were given to refit and again absorb a mass of reinforcements, most of them anti-aircraft gunners, virtually untrained and with no battle experience. The breakout from Anzio having developed into a fierce prolonged struggle, the hastily reformed Green Howards were again flung into the battle. At Ardea, on 31 May, they gallantly captured two hilltops but were then almost overrun by tanks, against which their only defence were a few PIATs. In some near miraculous way, the few remaining old hands staved off complete disaster. Regimental *esprit-de-corps* had done its work. It was the Battalion's last action in Italy; at the end of June 5th Division embarked for Alexandria to rebuild itself once more.

There is a postscript. Disobeying orders, a jeep-load of three officers entered Rome on 5 June at the same time as the American forces. They were not the first Green Howards to do so. Major D'Arcy Mander, lately second

Left: Private Mornington of the 1st Battalion, who has stepped on a mine at the Moletta crossing on 23 May 1944, is carried away by German prisoners.

Right: Preparing for 'D' Day. HM King George VI watching a Green Howard sergeant repairing a broken track on a Bren gun carrier in the United Kingdom on 23 February 1944.

in command of the 4th Battalion, had been there since January. Escaping from his prisoner-of-war camp after being taken at Gazala, he had made his way to Rome. Working with the underground Resistance, he was to earn a well-merited DSO.[11]

★★★

A wet misty morning enshrouding a bomb-shattered city greeted the 6th and 7th Battalions when they docked at Liverpool on 5 November 1943. Welcoming bands were absent. Much understrength, the two units were then further depleted as individuals left for instructional jobs. Some were undoubtedly battle-weary: courage is an expendable commodity and much had been spent in the Western Desert and in Sicily. Left in the two Battalions were small hard cores of experienced officers and men ready to tackle the job of building new fighting teams from a mass of young soldiers, few of them Green Howards or even Yorkshiremen. Among those lost to 69 Brigade was its own Brigadier, 'Red Ted' Cooke-Collis, sent to lend his knowledge, determination and leadership to another as yet unblooded infantry brigade.[12]

This inept handling of reinforcements was nothing new. (A draft of 200 Cameron Highlanders had inexplicably joined the 6th Battalion after Dunkirk; needless to say, the few survivors after three years had become in every way Green Howards, occupying key positions.) Overseas there could be problems in bringing depleted units up to strength, but it was disgraceful that in England Green Howard reinforcements could not be found. Rumour had it that the responsible individual in the War Office was unaware that the Green Howards had anything to do with Yorkshire.[13] Staff work in this war was much better than in the previous one, but a consistent failure was the posting of infantrymen to their own or like-minded regiments.

The original plans for the invasion of north-west Europe had 50th Division landing on the third day as a follow-up formation, but Montgomery's arrival changed everything. Intent on spear-heading the invasion with battle-tried formations, he switched the Division to the initial assault. As the commanding officer of the 7th Battalion recorded, the news did not improve morale. Although it was a compliment to the Division's great fighting qualities, experienced soldiers well knew what it would mean. For all that, most faced this fresh challenge with their usual dogged determination.[14]

***

Among the infantry of five divisions – British, Canadian and American – that stormed ashore over the Normandy beaches on the morning of 6 June 1944 was the 6th Green Howards, landing in the leading assault wave. Immediately behind was the 7th Battalion, 69 Brigade's reserve.[15]

A vivid and abiding memory for some of those Green Howards was that dawn scramble down nets hung over the ships' sides into storm-tossed landing-craft, burdened as they were with weapons, ammunition, shovels, respirators and all the rest of an infantryman's unavoidable impedimenta. As the craft formed up in the heavy sea, salvoes from large-calibre naval guns roared overhead, clouds of Typhoons swooped on the beaches, pounded by rocket ships as well as tanks and 25-pounders firing from the craft bringing them to the shore. For many, the arrival on the beaches was quite a welcome relief from sea-sickness.

For the German defenders to survive such fire seemed impossible, but as the leading companies of the 6th Green Howards and 5th East Yorkshire

Below: A wounded sergeant of the 7th Battalion in Normandy is lifted on to a Bren gun carrier on 17 June 1944.

waded the last sixty yards ashore, heavy machine-gun and mortar fire cut into them. With the help of tanks of 4th/7th Dragoon Guards, however, those Yorkshiremen seized their objectives. Prominent in the 6th Battalion was Sergeant-Major Stanley Hollis, who single handed stormed a by-passed pillbox that threatened his company from the rear. It was but the first of the series of heroic exploits that won him the the only Victoria Cross to be awarded for the D-Day assault.[16]

Two hours after landing, both 6th and 7th Battalions were a mile inland. By nightfall they had fought their way forward for another six miles, farther than any other British or American unit. In what has been described as the greatest military operation in history, those two veteran units had done supremely well, as had the rest of that fine 50th Division. To have failed to make good the bridgehead would have done the Allied operation irreparable harm, probably postponing the war's end by several years. It had cost the 6th Battalion less than 100 men, including two company commanders. It could have been far worse.

Major Macdonald Hull, second in command of the 6th, remembered how congratulations poured in from every direction. Never before or since had he been so proud of being a Green Howard.[17] And this young wartime officer had fought in every battle since Gazala; already he had won a bar to his Military Cross (it says something about the parsimony with which decorations were then awarded that he was the sole officer in the Regiment so to do).

★★★

At nightfall 50th Division, as part of XXX Corps, was almost on its planned objective, the Bayeux–Caen road; on its left the neighbouring corps had just failed to reach Caen. But the next day the full force of the *Panzer* reserves began to be felt as they started to hem in and delay the expansion of the fifty-mile-wide Allied bridgehead. It was now a race between the German armour, mauled all the time by air forces some 10,000 strong, and the build-up of fresh Allied formations and supplies across open beaches and through the artificial 'Mulberry' harbours in steadily worsening weather. In front of Villers Bocage, some twenty miles inland, 50th Division was to struggle for the next two months, now attacking, now staving off the German armour. This was attritional warfare similar in intensity to that of 1914-18, and in no way less bitter than that already undergone by the Green Howards at Alamein, Mareth, Akarit and Primasole Bridge. But in this thick Normandy bocage it was a different kind of war, one of movement in which infantrymen often fought tanks among the steep hills, deep valleys and small fields, bounded by near impenetrable hedged banks, all perfect cover for the concealed mortar and the sniper.

The 6th Battalion soon experienced some of the worst of it. Five days after landing and supported by a complete regiment of armour and two of artillery, it attacked towards Tilly-sur-Seulles, one of many such operations designed to pin the enemy down while Montgomery thrust hard towards Caen.

Information about the Germans facing the 6th Battalion was sparse, nor could the gunners spot their well-hidden targets. The supporting tanks were easy prey for concealed German bazookas and guns. Machine-guns swathed down the Green Howard riflemen as they struggled through the often neck-high crops; snipers dealt with junior officers and NCOs, unavoidably obvious as they exposed themselves while trying to do their job. Then, when the Battalion was almost on its objective, German tanks cut in behind the forward companies. Lieutenant-Colonel Robert Hastings, a Rifle Brigade officer who had commanded the Battalion superbly since its last weeks in Sicily, exercised the discretion the brigade commander had given him to withdraw. This was done in good order, but the day's fighting had cost the 6th more than two hundred casualties; a high proportion of them were those junior leaders.[18]

The pattern did not change. Replacements for losses were only rarely Green Howards. Many seemed to be ex-storemen, mess waiters or ration clerks; so short was the army now of junior officers that Canadians, Australians and Norwegians could be found among the platoon commanders. (In the first nineteen days in Normandy, the 6th Battalion lost twenty-nine officers and every rifle company sergeant but one.) Nevertheless, as ever, these heterogeneous reinforcements in their turn quite quickly became Green Howards. An infantryman had little but his self-respect, the presence of his friends and his pride of regiment to sustain him at such times when a not too serious wound seemed the sole alternative to death or permanent maiming. It was important to be a Green Howard.

Much of this fighting in the deadly *bocage*, occurred around Tilly-sur-Seulles, afterwards a Green Howard Battle Honour. Meanwhile vast tank battles had waged in the more open country around Caen; at the same time

Left: General Montgomery decorates CSM (later RSM) G. Calvert of the 6th Battalion with the ribbon of the DCM on 17 July 1944.

Right: Soldiers of the 6th Batalion cementing the 'Entente Cordiale' in Normandy.

the Americans had slowly ground their way through similar hedgerow country to capture Cherbourg and afterwards edge south into Brittany. Then, eight weeks after the landings and helped by the incessant British pressure on their left, in which the Green Howards had played such a very gallant part, the Americans broke out of the bridgehead to swing east towards Paris. Caught between the US armour and the British pressing south, and continually bombarded from the air, some dozen German divisions, most of high quality, were trapped and slaughtered around Falaise. Not until 16 August did Hitler permit the few survivors to withdraw through the almost closed Allied net.

In this final victorious battle the two Green Howards Battalions were again to the forefront. From 9 to 13 August, in yet more savage fighting, they threw the Germans out of St-Pierre-la-Vieille, an obscure village whose name was to be commemorated by a further Battle Honour.

★★★

When they saw the holocaust of the Falaise Gap, with roads choked by the carnage of men, horses and vehicles, the few weary and strained Green Howard survivors of the D-Day landings had every reason to think that the war was all but over. As the shattered remnant of the German armies fled towards the Seine, British 50th Division, as part of XXX Corps, took up the pursuit. In three weeks the 6th and 7th Battalions drove 500 miles in their dusty vehicles through a now liberated Europe, welcomed sometimes by the delirious French and Belgians with wine, fruit, bouquets, cheers, flag waving and kisses – sometimes by the bullets and bombs of German rearguards. On 1 September they crossed the Somme; the next day they passed through Arras. On 3 September Guards Armoured Division followed by 151 Brigade, entered Brussels to scenes of wild jubilation.

Above: Men of the 7th Battalion passing a knocked out half-track of
the 7th Armoured Division near Tracy Bocage, 4 August 1944.

But that exhilarating chase ended. The war was far from over. Along
the Albert Canal, the last obstacle before the Dutch frontier, the Germans
at last made a stand. On the right Guards Armoured Division had forced a
crossing over the Canal, but a second route was needed if the momentum of
the advance were to be kept up. Speed was still of the essence and to 50th
Division fell the task of gaining another bridgehead in the Gheel area.

The spearhead was to be the 6th Green Howards. Only while it was
being rushed forward the fifty miles to the Canal on 7 September, did its
commanding officer have the chance of seeing the ground and making his
plans.[19] The bridges had been blown; only two fourteen-man assault boats
and a few two-man reconnaissance boats were available. On the far bank
enemy positions were clearly visible. It all seemed hopeless, but the
brigadier's firm response to the colonel's protest was 'not to belly-ache'.

At 0100 hours in bright moonlight and total silence the two assault craft
began gingerly to edge across the Canal. Within an hour two companies of
the 6th Battalion had crossed and one was nearing its objective, 1,000 yards
to the south-east of the site chosen for a new bridge. Using the same boats,
one now leaking badly, the 7th Battalion also managed to cross virtually
unscathed. By then, however, the battle had begun and the 6th Battalion
was struggling to make good its initial foothold. And thus, for the next four
days, 69 Brigade and then 151 Brigade clung to and then expanded the
bridgehead against infantry and tanks, its enemy at first hard-fighting
Luftwaffe regiments but later the 2nd Parachute Division, last met at
Primasole Bridge. Not until the evening of 12 September were the Green
Howards relieved and the Germans seen to start to withdraw. Losses had
again been heavy. Gheel was to be another Battle Honour.

***

Montgomery now launched what was to have been the decisive blow in the west – *Operation 'Market Garden'*. Three airborne divisions were to seize crossings over the Rivers Maas, Waal and Neder Rijn. Then, up this sixty-mile-long, pencil-like corridor XXX Corps was to swoop, passing through Eindhoven and Nijmegen to reach British 1st Airborne Division, holding their bridgehead at Arnhem over the Neder Rijn, forty-eight hours after crossing the start-line. A right hook would then encircle the Ruhr. It was an overbold attempt to end the war in 1944 and thus save Europe further misery, and it was to fail.

The 6th and 7th Green Howards were not drawn into the main battle until 25 September, only hours before the pitiful remnant of 1st Airborne was withdrawn over the river. Brought up from reserve to the so-called 'Island', the flat low-lying polderland between the Waal and the Neder Rijn, the two Battalions made good in hard fighting a section of the corridor just north-east of Nijmegen, the area in which the Guards Armoured Division had been finally halted. Here on 27 September the 21-year-old Captain John Franklyn, only son of General Harold Franklyn, was killed commanding a company of the 6th Battalion. He had already been wounded – in Normandy ten days after the landing. Son and grandson of Colonels of the Regiment, his death was a grievous loss to the Green Howards.

This fighting was but a prelude. The withdrawal from Arnhem had allowed the enemy to mass its forces against the east flank of the 'Island'. At dawn on 1 October, in thick mist, a major counter-attack hit 7th Battalion's sector. For eighteen hours, until at last relieved by 5th East Yorkshires, the Green Howards held their slit-trenches against aggressive infantry backed by large numbers of the formidable Tiger and Panther tanks. One private soldier, who had joined only three days before, won the DCM by knocking out one Panther with his PIAT from point-blank range and then hitting and driving off a second. The over-used word 'epic' well describes what was to be the last major action of that Territorial Battalion.

For 'Neder Rijn' was to be its last Battle Honour. Enduring the mud, rain and bitter winds of that autumn for two more months, the 6th and 7th battled on in the 'Island' in what had become, for the time being, a static war. Then, at the end of November, they heard that their fighting career and that of 50th Division was over. Britain was running short of men. No longer could the TT Division be kept up to strength.

Cadres from the 6th and 7th Green Howards returned to Yorkshire to pass on their infantry skills to men from other arms. The two Battalions and their sister units of the East Yorkshire Regiment and the Durham Light Infantry had seen more fighting than any other British infantry. It was a magnificent record, never to be forgotten.

***

The Western Front had not seen the last of the Green Howards. On 30 April 1945 Hitler committed suicide in his Berlin bunker, four days later the German forces in north-west Germany, Holland and Denmark surrendered to Field Marshal Montgomery. On the very day of Hitler's death, the 1st Battalion was back in action, crossing the River Elbe, successfully assaulted in the early hours of that morning.

The 5th Division, after six months' hard training in Palestine, was once again a first-class fighting formation. Twice it received orders to return to Eighth Army, still fighting its way up the length of Italy. The second time it actually reached Italy, only to be switched to France and then moved up through Belgium to Germany.

So it was that Green Howards saw action again in the very final stage of the north-west Europe campaign, fighting against the few fanatical young Nazis still resisting with the tenacity of desperation. That last minor battle on 1 May was to cost the Green Howards dear. Men died who had been with the Battalion since Norway. CSM Peacock, who had won that MM in Palestine six years before, as well as a Norwegian Military Cross at Otta, was to be awarded the DCM for his gallantry and skill in command of his company north of the Elbe after every officer had been hit. There was a special poignancy about those deaths, incurred when the war was so near its end. But there was a sound reason for them. In those last few days Montgomery was racing for the Baltic so as to forestall the Russians, his object to save Denmark from the Red Army and ensure that any campaign to liberate Norway should be Anglo-American. Scandinavia had reason to be grateful to those British soldiers.

Among the seriously wounded was a young wartime officer, Major Peter Howell. With commendable understatement, he found it 'rather trying' to have been so badly hit so close to the war's end.[20] His name is mentioned because his battle experience among Green Howards was matched only by Major Macdonald Hull. Howell had already distinguished himself with the 4th Battalion in front of Gazala, the 6th at First Alamein and with the 1st at Anzio; like so many others, he was never even decorated. His successful subsequent career in commerce he attributed largely to what he had learned among Green Howards: the importance of decisive leadership, caring for those for whom he was responsible, together with the vital importance of sound morale, exemplified by the Regiment.

★★★

Horrible though this war had been, British losses numbered only a quarter of those killed and wounded from 1914 to 1918. Likewise Green Howard losses had been appreciably less. Although in battles such as Anzio, Alamein and Normandy the fighting had been just as intense and losses as great, life for the soldier was rather more varied, relief from war's squalor was more often found and the small comforts of life more frequently enjoyed. Above all, it had been a more professional war. Modern weapons, transport and communications had made for quick moving battles demanding highly skilled infantry. It was fortunate that the regular cadres were not again wiped out in the first few months as in 1914. Their expertise had been preserved to produce the type of soldier needed.

★★★

A postscript to the story of the Second World War is the alliance it produced in January 1948 with the Rocky Mountain Rangers – a Canadian militia regiment. In the spring of 1944 Green Howards had helped settle the newly arrived Canadians into their camp in the North Yorkshire Moors and from it stemmed a long-lasting friendship. First raised in 1885 as a light cavalry contingent of farmers and ranchers during the Northwest Rebellion of 1885, Rangers have fought in every military action in which Canada has been involved. Based on Kamloops in British Columbia, their motto is 'Kloshe Nanitch', the local Indian for 'Keep a Good Lookout'.

But even older is the alliance with the Queen's York Rangers (1st American Regiment), Royal Canadian Armoured Corps, which dates from 1927. Also a regiment of militia, it is an amalgamation of the Queen's Rangers and the York Rangers. The former unit has served continuously since 1866, but its ancestry can be traced back to 1756, the year that Robert Rogers raised the famous Rogers' Rangers for service against France in the Seven Years War. During the American Revolution, Rogers formed a new unit of scouts which came to be known as the Queen's Rangers and was granted the title '1st American Regiment' by King George III. The Rangers were the sole unit to save their Colours from surrender at Yorktown, and these can still be seen at Fort York Armoury in Toronto. Today the regiment has the role of medium reconnaissance, scouts for the field army as were Rogers' Rangers.

Since the Second World War the Regiment has also developed a close and happy unofficial affiliation with the Royal Bodyguard of the Norwegian

Right: The Regimental War Memorial to the dead of two World Wars in Richmond.

monarchs, H.M. *Kongens Garde* (the King's Guard) of Norway. Exchange visits between the two Regiments have been frequent and the King's Guard have provided unstinted, friendly and generous help to the many Green Howards, who as individuals or in formed bodies have been lucky enough to visit Norway.

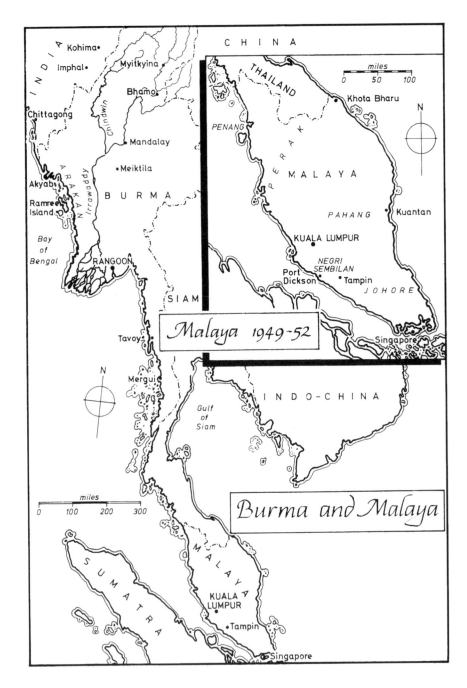

11. RETREAT FROM EMPIRE, AND THE NATIONAL SERVICEMAN, 1945-68

In the two decades that followed Japan's capitulation, brought about when successive atomic bombs destroyed Hiroshima and Nagasaki and their inhabitants in August 1945, more than 7,000 British soldiers were to be killed or wounded in different corners of the world.[1] Only one year passed in which a British soldier did not die in action. A high proportion, possibly over half, were National Servicemen; among officers alone, some 250 were killed.

***

Some of the work was old-fashioned 'peace-keeping', a task that fell to 2nd Green Howards during 1946 in Calcutta at the time of the communal massacres that marked the bloody run-up to Indian independence in which a million may have died. On one occasion, for five successive days in the clammy monsoon heat, young Green Howards (anyone with forty months' service overseas had left for home) did what they could in the stench and smoke to keep the murderous mobs from their screaming victims. With a handful of other units they eventually restored some semblance of peace to the streets of that enormous city, littered by then with 4,000 rotting corpses. In the absence of anyone else the British infantry had the job of removing the ghastly remains. Officers and others volunteered to drive the rubbish carts into which Indian sweepers dumped them. A staff officer said of those volunteers, 'They drove their trucks until they were almost in a stupor; they stank so much that even their own men recoiled from them . . . they did not eat for two days but lived on cigarettes and neat gin'.[2]

A few months later, after twenty-two years overseas, the 2nd Battalion took ship for Port Sudan and from there moved to Khartoum, in a few years to be the capital of an independent Sudan. There, on 31 December 1948, the Battalion disappeared, in name but not in body, into a limbo described in the weird contemporary jargon, as 'suspended animation'. To avoid individuals being moved around unnecessarily, the 1st Battalion, back in England from Germany, was broken up. Its name, property and colours were transferred to the 2nd Battalion which in effect became the 1st.

So once again, the 2nd Battalion had disappeared, a sad loss suffered by all line regiments, the structure of the post-war army demanding a better balance between the teeth arms and their supporting administrative services. To reduce training staffs and ease posting problems, infantry regiments were then grouped by counties. After seventy-three years recruits were no longer to be trained at the Richmond Depot; the Green Howards were to merge some of their loyalties into a larger body, the Yorkshire and Northumbrian Brigade Group, its members their fellow Yorkshiremen of the East Yorkshire Regiment, the West Yorkshire Regiment, the Duke of Wellington's Regiment and the York and Lancaster Regiment, as well as the 'Geordies' of the Royal Northumberland Fusiliers. A common background and close wartime associations ensured that there was no friction. Such was the inter-posting of officers that in 1950 only three of thirteen eligible majors were serving with the 1st Battalion in Malaya; of thirty captains, no more than five were there.[3] At one time the Adjutant was an East Yorkshireman and the Adjutant of the 1st East Yorkshires was a Green Howard.

***

The move to Malaya in 1949 had come about through the urgent need of extra troops to cope with the Communist insurrection there. This had erupted the previous year, the first major conflagration with which Britain had to cope as she began to lay aside her imperial responsibilities, an inevitable task but one she strove to discharge in an orderly manner.[4] Prior to the Japanese invasion in 1941 Malaya had been a well-governed country, prosperous from its rubber and tin, both worked largely by immigrant Chinese and South Indian labourers. A few Communist cells had sprung up among the Chinese community in the pre-war years, a far from widespread movement but one that provided the inspiration and organization for the small force of guerrillas which resisted the Japanese occupation. In part British trained and led, towards the end of the war these Communist Chinese were lavishly supplied with the weapons which they would soon turn against their donors.

Post-war instability provided the right conditions for a revolt. India and Burma had been granted their independence, there was near chaos in the Dutch East Indies and the bitter conflict that led to the birth of Vietnam was already under way in the French colonies in Indo-China. Lending encourage-ment and often material help was the Soviet Union, aggressively expansionist and bedevilled by unreasonable suspicion of her wartime allies

What followed was carefully planned, a classic four-stage Communist takeover. First the police, government officials and the handful of European miners and planters would be driven to seek refuge in the towns. Then in these country areas so 'liberated', an insurgent army would be assembled. In the third stage these 'liberated areas' would be expanded as the economy and security collapsed, following which the insurgent army would take the field in open warfare against the British troops.

Inspired by the Communist-organized Calcutta Youth Festival, held in February 1948 and attended by a large Russian delegation, the erstwhile wartime fighters dug up their weapons and moved back into their jungle camps, often not too enthusiastically. In April sabotage and intimidation began with the killing of Chinese, Tamil and Malay workers in the mines and on the plantations. The murder of five Europeans in June signalled the start of a major Communist uprising that would take twelve years to defeat

Above: On 13 December 1946, General Sir Harold Franklyn unveiled the new Royal Scot class locomotive No. 6133 'The Green Howards'. (by courtesy of the National Railway Museum, York)

Right: In 1963 a subsequent Colonel of the Regiment, Brigadier G. W. Eden, accepts the name-plate of the Royal Scot class locomotive when he in turn unveiled the Deltic class diesel which carried on the name. To the left of Brigadier Eden, wearing a bowler hat, is Colonel J. M. Forbes, the Regimental Secretary.

and see the creation of an independent and democratic Federation of Malaya.

When the Green Howards moved into Pahang in September 1949 it was already clear that the Communists would not win popular support, particularly from the Malay majority of the population. Although the active guerrillas, who were to describe themselves, first and inaccurately, as the Malayan People's Anti-British Army, and later as the Malayan Races Anti-British Army, usually numbered no more than 4,000 at any one time, they operated in perfect guerrilla country, sparsely inhabited – four-fifths of it thick tropical jungle astride a spinal mountain range.

As so often happens with a unit fresh to anti-guerrilla fighting, the Green Howards at first had little success in their unfamiliar surroundings. A subaltern at the time, Nigel Bagnall, later Field Marshal and Chief of the General Staff, remembered how, 'Nobody else had any real idea as to what we were supposed to be doing either. They had all acquired some basic jungle skills but the operational situation was, to put it mildly, unclear . . .'[5] The police had mostly barricaded themselves into their stations, roads were open to ambush; weapons were unsuitable for such a campaign until American carbines and Australian Owen guns arrived in place of single-shot .303in Mk V rifles and cumbersome Stens.

Patrols, sometimes a company strong, and rarely less than a platoon, toiled through the jungle, for as long as fourteen days at a stretch, seeking a wary and elusive enemy about whom little as yet was known. Contacts, when made, were usually a matter of chance. During the first three months Green Howards met the Chinese only five times, killing one of them at the cost of one soldier. Just before Christmas another Green Howard died, together with an aboriginal Sakai tracker, and two soldiers were wounded in an unexpected encounter in the jungle half-light with a patrol of Gurkhas; three months later yet another Green Howard was killed when two patrols from the same company clashed. As always in war, such accidents happen, as tragic as they are unavoidable.

Meanwhile, the Communists were creating some twenty incidents each day up and down the country; in a single month they killed 200 people, most of them police or innocent civilians. Nevertheless the security forces were just holding their own. The isolated Europeans – government officials, miners or planters – stayed put, as did most of their wives, despite the ever-present danger. The police, almost entirely Malay and demoralized after the Japanese occupation, were being reinforced and given fresh European leadership. Because few of the known Chinese Communists possessed local citizenship, numbers of them were being deported to their homeland, a salutary weapon indeed.

Below: The officers and sergeants of 'C' Company of the 1st Battalion in Malaya in 1950. Front Row: CSM Lord, Maj G. Ritchie, Col Sgt Samme. Behind: 2/Lt Tyzack, Col Sgt Philips, ?, 2/Lt Sabine, Sgt White, Sgt Snaith, 2/Lt Cooley.

In March 1950 the Battalion was moved from Pahang to the adjoining and more northerly state of Negri Sembilan. By now its skills were improving fast. Young officers and NCOs, both regulars and National Servicemen, were developing into fine leaders; high standards of jungle-craft, marksmanship and fitness were being inculcated, qualities essential to offset the hardships and frequent disappointments of a life both mentally and physically exhausting. They learned to move quietly and listen for enemy noises. Patrols were often reduced in size to three or four individuals, making for stealth and alertness. Men learned to lie in ambush on jungle tracks for days and nights on end. From Malays and the cheerful headhunting Dyak and Iban trackers brought in from Sarawak, lads of nineteen or twenty, fresh off the streets of Middlesbrough, learned the skills of detecting traces of enemy movement in the 'ulu'.

As Major J.B. Oldfield, a company commander and later author of *The Green Howards in Malaya*, put it, although war is often described as long periods of intense boredom punctuated by moments of intense fear, the Malayan fighting could well be defined as 'a long period of unceasing effort punctuated by moments of intense activity and rare success'.[6] And about half the private soldiers and some junior NCOs were National Servicemen, while most of the regulars were equally young, recruited on three-year engagements.

The enemy were both skilful and courageous. No sooner had the Battalion arrived in Negri Sembilan than a three-vehicle convoy drove into a cleverly laid ambush, manned by fifty to seventy terrorists, overlooking from high embankments a road that had been mined in seven separate places. Three Green Howards died and another was wounded, but it could have been far worse. By now, however, the smaller patrols, working just inside or on the edges of the jungle, were gaining some successes and the score of terrorists killed was starting to mount. But senior commanders were still wasting far too much effort on what were virtually large-scale partridge drives that could involve half a dozen battalions at a time, tactics that had so often failed in the past. So much of what the British Army had learned in anti-guerrilla campaigns fought on terrain ranging from North America to South Africa had been forgotten.

The arrival in April 1950 of Lieutenant-General Sir Harold Briggs as Director of Operations, his brief to co-ordinate the operations of all departments, military and civil, had given the campaign fresh impetus. The 'Briggs Plan', put into effect in June, was aimed at denying to the insurgents the supplies which they extracted from the half million or so Chinese squatters who lived around the jungle's edge. By resettling these squatters into villages, it was possible to give them a large measure of protection against terrorist demands and atrocities. As the Communist grip on them gradually slackened, the villagers discovered that it was safe to talk and the insurgents found it hard to feed and clothe themselves.

At the same time, liaison among the security forces was placed upon a sound footing by Joint Local Committees, upon which the local district officer, policeman and army company commander worked closely together.

The latter, obtaining his information at first hand, was in future allowed to control his own platoons and largely run his own operational area. At the same time intelligence improved and jungle contacts depended far less on chance sightings.

A spell in Singapore had been planned for Christmas 1950 during which the Battalion could absorb reinforcements, rest and retrain, and the married men see a little of the families, so near and yet so very far away. The unexpected outbreak of serious Muslim rioting in a city completely devoid of combatant units led, however, to the Green Howards and three Gurkha battalions being rushed out of the jungle and over the causeway into Singapore, some still with the jungle mud on them. But by the time they arrived the base troops had coped admirably and the trouble was almost over.

By Christmas the Green Howards could turn their attention to their families, to the traditional seasonal celebrations, and to recapturing the precision that enabled them to Troop the Colour on a much postponed Alma Day parade. In the words of the Malayan Broadcasting Corporation commentator 'not even the Brigade of Guards could have done it better'.[7] Photographs bear this out. And a high proportion of those marching past so superbly in line were National Servicemen. But there was to be little time for that hoped for rest. Before the Battalion moved back up country to North Johore in February, much thought and energy was to be spent on training for the next spell in action.

The full rewards came when the Battalion was switched to the Tampin area in April 1951. There they were to remain for fifteen months, time to allow them to come to terms with the 'local regiment' of the Malayan Races Liberation Army and the country in which it operated. Bagnall, who had already won an MC in Pahang and was to add a bar to it in Tampin, remembered how many of the junior leaders, both officers and NCOs, upon whom success largely depended, had perfected their tactics and developed an eye for the country. Sometimes help came from guerrillas who had been captured or surrendered (Surrendered Enemy Personnel or SEP was the jargon). The first had given himself up to the Battalion in October 1951 and straight away agreed to guide the troops back to his former comrades. Such behaviour was a source of wonder to the soldiers. Disillusioned by failure, discouraged by the death of friends, worn-out by years in the jungle and harried by the army, once an insurgent gave up the struggle, he seemed to have no shame in betraying his associates. Some stayed with the Battalion, developing into something approaching useful mascots.

In July 1952 the Battalion moved north to Perak for the last ten weeks of its tour, after all but eliminating the terrorists around Tampin. In all sixty-five had been killed or captured in forty separate actions. A further nineteen were to be eliminated in Perak, bringing the total for the tour to more than 100. One company commander and eight soldiers had been killed by the enemy; eleven had died from other causes; no record seems to have been kept of the number wounded.

In thirty-eight months, seventy-one officers and 1,646 other ranks had served with the Battalion. Only one officer and seventy-five of those who

Above: Richmond Castle, by David Muirhead.

Below: Patrol in the Ardoyne in 1971. Painted by David Shepherd. *(Reproduced by courtesy of the artist)*

Left: *Oil painting by Terence Cuneo of the Regimental Chapel at the unveiling on 3 June 1959 of the memorial to His Majesty King Haakon VII by his son, King Olav V, his successor as Colonel-in-Chief. On the right is Major-General A. E. Robinson, the Colonel of the Regiment, and in the foreground is His Grace the Archbishop of York, Dr. A. M. Ramsey. (Reproduced by courtesy of the artist)*

Below: *HM King Olav V, the Colonel-in-Chief, inspects the 1st Battalion at Catterick on 24 June 1989 at the celebration of the Regiment's Tercentenary, one year late because of operations in Northern Ireland. On his left is the Commanding Officer, Lieutenant-Colonel J. S. W. Powell, and immediately behind is the Colonel of the Regiment, Lieutenant-General Sir Peter Inge.*

Top: *A group photograph of the 1st Battalion wive's club in Malaya. Because it was largely a National Service unit, only a small number of families were present.*

Above: *The Archbishop of York, Dr. Garbutt, visits the 1st Battalion in Malaya on 14 January 1952.*

Right: *The branch secretary of the Tampin Min Yuen who surrendered to the 1st Battalion in Malaya on 11 June 1952. Like many of his fellows, he willingly led a patrol back to attack his former comrades.*

Above: Insurgent casualties; photograph taken in 1950 or 1951.

Below: When one insurgent camp was attacked, a camera was found complete with film. When developed, this posed photograph of a group of Chinese communists was revealed.

had arrived with it were there at the end of the tour. Not only had there been the normal turnover but large drafts had also been found to reinforce the Royal Northumberland Fusiliers, fighting a more conventional and more bloody war against the Chinese in Korea. It was extraordinary that such a high level of professionalism had been achieved among the ever changing faces.

John Oldfield, later Colonel of the Regiment, in his account of the campaign was careful not 'to do ourselves too much honour'.[8] An impartial commentator was less inhibited. The latter declared that 'the Green Howards departed, heavily laden with tribute', adding that they would blush to be so singled out. Nevertheless he did do so. They had, he said, 'prospered from the confidence they built up among the locals, they infused it by the success they gained, and this in turn stemmed from a continual urge for self-improvement . . .' Yorkshire tenacity, he wrote, had been harnessed to opportunism.[9]

General Sir Gerald Templer, who had arrived in Malaya as High Commisioner on 7 February 1952, the day after the early death of King George VI and the accession of his youthful daughter as Queen Elizabeth II, was to break the backbone of the insurrection. He was a man who wasted few words, none on unnecessary compliments. At his farewell to those Green Howards he said:

> I am not going to tell you that you have made a name for yourselves in Malaya. You already had a name before you arrived. What you have done has been to add to it.[10]

★★★

On its arrival home in December 1952,[11] a hutted camp at Barnard Castle and a bitter Durham winter awaited the 1st Battalion. There too they found the 2nd Battalion, reformed again after its demise in 1949, the Korean War having aggravated the Army's lack of infantry, a commodity always in short supply when trouble occurs. More battalions were urgently needed and the Regiment, because of its excellent recruiting, was one of the eight selected for expansion. The previous year the depot at Richmond had again been set to its primary task of training the Regiment's recruits.

With the two regular Battalions alongside each other, the only occasion this had ever happened, the time was perfect to exercise the Regiment's privilege of marching both its regular and Territorial units through Beverley, Bridlington, Middlesbrough, Redcar, Richmond and Scarborough, the six Yorkshire boroughs that have, over the years, honoured the Regiment with their Freedom.

★★★

In the spring of 1953 the 1st Battalion was away again on its travels, this time to Graz in Austria, still occupied by the erstwhile wartime allies, Britain, France, the United States and the Soviet Union – the first three all wary of leaving the country until the Russians had gone as well. Friendly in the immediate post-war years, relieved to be rid of their Nazi rulers, by 1953 the Austrians were tired of having those foreign troops about the place. As a result the atmosphere was not altogether pleasant. A shortage of officers and men also made life difficult; with the other six battalions of the York and Northumberland Brigade all actively employed, either in Korea, Malaya, Cyprus or Aden, the 1st Green Howards was starved of reinforcements. So it was with some relief that, on the reduction of the Austrian garrison from three battalions to one, the Green Howards moved to the British Army of the Rhine (BAOR) to join 61 Lorried Infantry Brigade in 6th Armoured Division at Minden. Life there was far more of a challenge to all ranks of the Battalion.

After 1945 the army's task was not merely to keep order in a steadily contracting empire, but also to help guard western Europe from the clear likelihood of invasion from the east. Behind what Churchill had so graphically depicted as the 'Iron Curtain', Russian-dominated Communist dictatorships ruled the countries of eastern Europe, while in the west Communist Parties exploited battered societies struggling to reshape themselves. The Cold War was under way. The land blockade of the western occupation zones of Berlin, carrying with it the threat of a Soviet invasion of West Germany, led in 1949 to the creation of the North Atlantic Treaty Organization (NATO), the military machinery for defence against further aggression from the east. No longer an occupation force, the British troops in Germany were part of the west's defences. The early 1950s were, then, a time of feverish activity in BAOR, with divisional commanders tending to compete one with another. Training was hard but interesting. Soon the 1st Battalion began to receive reinforcements of proper quality.

One consequence of this renewed threat of war was the spread of a refreshing professionalism in the army. This had always been the mark of its NCOs; after the war, it was extended into the officer corps. Pay, although still poor, was just adequate: officers could now live on it. They were, as a result, drawn from a wider field. Without as yet being overly ambitious, it was now the rule rather than an exception for them to take a proper interest in their jobs. In 1939 only two serving Green Howards had qualified at the Staff College; as with many other regiments, it had not been quite the thing to do. After 1945 most officers sought a place at Camberley; many succeeded. But one drawback was that it became all but impossible to achieve command without those magic initials 'psc'. After 1945 only one Green Howard was to command the 1st or 2nd Battalions without having been staff trained. The consequence was that some excellent regimental officers left the army when they failed Camberley entry because they knew that their chances of commanding their battalion had gone.

★★★

A year after the re-raising of the 2nd Battalion had been announced, it sailed for the Suez Canal.[12] Life there was in every way arid. In 1948 the thankless task of keeping Jews and Arab from one another's throats in Palestine had been abandoned. The previous year all British troops still in Egypt had been concentrated in the Suez Canal Zone. There a vast base was expensively constructed for the purpose of keeping open communications with an eastern empire in the process of rapid dissolution. Harassment of the troops from increasingly active Egyptian terrorists brought the British Government to the negotiating table in July 1953, but for another year the 2nd Battalion was to divide its time between training and guards; confined to its Kabrit camp when not on exercises, its only relaxations were football, swimming and dinghy sailing on the Bitter Lakes, its married men tantalised by thoughts of their families just across the water at Larnaca in Cyprus. It says much for the calibre of the commanding officer and the experienced handful around whom the unit had been built that the 2nd Battalion was quickly acknowledged as being one of the best units in Egypt.

At last, in August 1954, the Battalion joined its families – by the unconventional method of practising an assault landing over beaches.

Above: A family regiment. Armistice Day Service held by the recently re-raised 2nd Battalion in Cyprus in 1954. The small boy to the left of the front row is a future commanding officer of the 1st Battalion.

Withdrawal from the Canal Zone had begun in earnest. A new base was to be established in Cyprus with GHQ, Middle East, transferring there as well. Until then the garrison of the island had consisted of a single infantry company.

It was a decision that aroused hostility among the Greek Cypriot majority, about four-fifths of the population, in many of whom had been implanted a fervent desire for *Enosis* – union with Greece. Archbishop Makarios, in providing its political leadership, was following the tradition of the Greek Orthodox clergy of eastern Europe. Promise of a new constitution granting a high degree of self-government only brought protesting crowds on to the streets of the capital, Nicosia.

For the time being, however, neither the Green Howards nor the 2nd Royal Inniskilling Fusiliers who arrived the following month, detected any overt hostility among either Greek or Turk. Despite the discomforts of a tented camp in a Dhekelia winter, they found the island enchanting, the open-hearted friendliness and hospitality of Greeks and Turks of all classes contributing much to the pleasure of life. Old men in the villages would offer hospitality in their coffee-house; to ask the way would result in being taken to one's destination. The only complaint voiced at Larnaca cocktail parties was about the Government's lack of tact in placing a British police officer fresh from West Africa, charming though he was, in charge locally; after all, they objected, they were Europeans.

From a soldiering standpoint it was an unsatisfactory period with companies employed building army warehouses under sapper supervision on the nearby airfield. But on 1 April 1955 the concluding hours of a company test exercise were disturbed by distant explosions. In a jocular way the company commander remarked to his sergeant-major, alongside him in a slit trench, 'So the Cypriots have risen'. A couple of hours later a despatch rider arrived with orders for an immediate return to camp.

★★★

In the background, Colonel Grivas, a regular Greek army officer, had been setting up the terrorist organisation, EOKA. The most unnecessary colonial conflict of the post-war years had begun.

At first the rather ineffective bombing seemed no more than an isolated outburst of violence, of comparatively small consequence, but suspicion and hostility rapidly spread: to be friendly with the British would bring unpleasant retribution. Green Howard sections had to be despatched to protect and bolster threatened police stations. Well-organized riots and demonstrations, usually by schoolchildren, proliferated. Detachments around the island multiplied. When the Greek members of the police force began to succumb to pressure, contrary to every current tenet of 'Duties in Aid of Civil Power' platoons were given instruction in the use of police batons and shields. Searches of villages, surrounded at dawn, produced little in the way of results. Soon civilians and policemen were being murdered; a bomb was thrown into the garden of a married quarter.

Below: Guard of Honour commanded by the first author and mounted in 1955 by the 2nd Battalion at Larnaca, Cyprus, for the departing Governor, Sir Robert Armitage. Nearly every member of this Guard was a National Serviceman, including Sergeant Smith, the left marker. Field Marshal Sir John Harding was to replace Sir Robert, and a few days later another Guard was mounted for him at less than twelve hours' notice.

Above: The 1955 Alma Day celebration by the 2nd Battalion in Cyprus by those members not engaged on operations. A 'Donkey Derby'. The second rider from the left is the future Field Marshal Sir Nigel Bagnall. On his left is the Quartermaster, the future Lieutenant-Colonel E. D. Sleight.

More troops began to arrive to relieve the now hard-pressed Green Howards and 'Skins'. Field Marshal Sir John Harding, on handing over as CIGS, was brought in as Governor. Landing by helicopter at 0730 hours on 13 October, to pay his first visit to Larnaca, he was met by a superb Green Howard guard of honour, nearly all National Servicemen, who had been woken and warned for the job exactly four hours before. Later, in the mess, Bagnall, by then a captain, mentioned to the new Governor that he had not joined the army to build warehouses. An officer, standing in the same group, observed Sir John's reaction and reflected that Bagnall's future prospects were poor. Harding was to live to see Sir Nigel Bagnall receive his field marshal's baton.

Fortunately perhaps, because those who had known the island before the bloodshed started found it difficult to dislike those pleasant Cypriots, the 2nd Battalion saw no more than the opening stages of what would develop into a bitter four-year campaign and lead eventually to the pointless partition of the island.

Once more, however, the army was contracting and those newly raised 2nd Battalions were to disappear. On 28 October 1955 the 1st Battalion the Middlesex Regiment arrived, landing by lighters that were to take the 2nd Battalion out to the waiting troopship. It was a far from orthodox relief. Stone-throwing schoolchildren ambushed vehicles bringing the troops down to the Famagusta quayside, access to which had to be cleared with batons. Television reporters and the Battalion's families watched or listened to it all.

Back again at Barnard Castle, in mid-March 1956 that fine 2nd Battalion once again vanished into the oblivion of 'suspended animation'.

★★★

No sooner was the 1st Battalion fully trained to operate effectively as part of an armoured division in the North German plains, than it was shipped out to the Far East,[13] a move that coincided with the regrettable disappearance of its sister regular battalion from the Army List. Its three-year tour in Hong Kong, a pleasant station in which tower blocks were only just starting to burgeon, was to provide a fresh and varied series of challenges for its National Servicemen, coping first with ugly riots, later a typhoon and finally a major part in a magnificent torchlight tattoo, possibly the finest such event ever organized on the island – something of a change from the ceremonial parades, guards and duties that inevitably fell to the lot of such garrisons.

One such parade, held on 19 August 1958, was to celebrate King Olav V of Norway's gracious assumption of the appointment of Colonel-in-Chief of the Regiment. In this he had followed his father, King Haakon VII, who had died the previous year. It was to be the start of a happy association that was further to strengthen the bonds between Scandinavia and the Regiment, bonds that King Haakon had done so much to consolidate.

Throughout his life King Olav's links with Britain had been strong. A Danish prince born in England on the royal estate at Sandringham in 1903, two years before Norway broke away from Sweden, in due course he was to study at Balliol College, Oxford. An able performer in a variety of sports, as a yachtsman he won Norway's first Olympic Gold Medal at the 1928 Games. After the German invasion of his country he was, with his father and his country's Cabinet, to endure two hazardous months; retreating ever northwards in front of the advancing Germans until eventually they were to be evacuated from Tromso in a British cruiser. In exile for the rest of the war, Crown Prince Olav and his father became the rallying point for their countrymen's determined struggle against Hitler's Germany, supporting and enthusing both the Resistance in Norway itself and the armed forces and mercantile marine of Free Norway working with the Western Allies.

King's Olav's death in January 1991 was to sadden the Regiment. His genuine interest is reflected in its close affiliation with the *Kongens Garde*, the Norwegian Household Troops, an affiliation that was far from a formality and one strengthened over the years by the succession of Green Howard officers who were to hold military posts in Norway. The King always seemed

Left: Hong Kong Military Pageant 1958. The leading figure is Major G. T. M. Scrope, a future Regimental Secretary, playing Colonel Francis Luttrell.

Right: Stanley Fort and Square, Hong Kong, where the 1st Battalion was stationed in 1958.

Right: Guard of Honour provided by the Regiment outside Holyrood House.

Right: His Majesty inspects the Guard. Its commander is Major G. T. M. Scrope. Behind him is the Colonel of the Regiment, Brigadier G. W. Eden.

to enjoy himself when visiting units of the Regiment; it was invariably a delight to listen to his off-the-cuff and noteless dinner speeches delivered in perfect English. He was much more than a figurehead to the Green Howards.

★★★

Another three-and-a-half year stint in Germany, stationed this time in Iserlohn, followed 1st Battalion's departure from Hong Kong in mid-1959. Half-way through this tour its drivers relinquished their 3-ton troop carrying trucks and learned the art of handling 1-ton armoured personnel carriers, a significant step forward in the role of the infantry soldier. Sport and training were the order of the day, but on two occasions the Battalion sent home Guards of Honour. The first, in June 1961, was for the arrival of HM The Queen at York Station when HRH The Duke of Kent married Miss Katharine Worsley, the daughter of that Green Howard captain of the Yorkshire Cricket Club, by then Lord Lieutenant of his County. The second was the House Guard at Holyrood in Edinburgh during the State Visit of its Colonel-in-Chief to that City.[14]

During those years in BAOR the army was to undergo yet another major upheaval. In 1957 it numbered 373,000, not counting the Gurkhas. Now it was to be halved, the infantry cut from seventy-seven to sixty battalions. Further amalgamations were the outcome, one of them between the East and the West Yorkshire Regiments to form the Prince of Wales's Own Regiment of Yorkshire. The loss of the East Yorkshire Regiment was especially sad for many Green Howards. In two wars the two Regiments had been brigaded together in both the celebrated 69 and 150 Brigades, as well as in several New Army brigades during the First World War. If a Green Howard could not serve in one of his own units, he could often make himself feel at home with the men of Hull and the East Riding.

At the same time the Royal Northumberland Fusiliers left the Yorkshire and Northumberland Brigade Group, then to be transformed into the Yorkshire Brigade with its headquarters and brigade depot at Strensall. Regimental cap badges were discarded for a common brigade badge of a Yorkshire Rose, an unpopular move with all concerned. To save manpower, recruit training was centralized at Strensall, a change that led in 1961 to the final closure of the Richmond Depot and to those familiar grey buildings on top of the hill becoming first an approved school and then a housing estate.

Needless to say it was with much relief at the Regiment's survival that the 1st Battalion left Germany in 1963 for a three-year tour in Libya, still, for a few short years, a British base in an independent Arab kingdom that would in 1969 become Colonel Gadaffi's Revolutionary Socialist Republic.

These cuts, that affected all three Services, had become possible by the contraction of the country's overseas commitments. A further factor was the decision that the country would depend primarily upon the nuclear deterrent to meet the threat from the Communist east, so allowing a reduction in conventional forces; in theory the strategic mobility to move troops quickly around the world was to come from a large force of transport aircraft.

★★★

Above: The 1st Battalion training in Tripoli in 1963.

The cuts led also, in 1960, to the abolition of conscription. Alone among European nations, Britain's armed forces would again become wholly volunteer. Once more the British would depend upon a small and highly trained professional force, a move welcomed in many circles as one that would relieve the army of that never ending chore of training successive intakes of conscripts. Much, however, was to be lost by this reversion to traditional practice. Again the army would lose touch with the civilian population, which in its turn would become increasingly ignorant of its army. Conscription had also brought into the services high-quality men, a cross-section of the nation. Those very bright young National Service subalterns, often high-spirited and at times hard to control, would often be remembered with affection and respect by Green Howards in the future. Remarkably high standards were achieved by the rank and file during their short period of service. It was commonplace to have section commanders who were National Service men, and they could perform superbly, whether training in an armoured division in north Germany or sweating it out in a Malayan jungle. In Cyprus 2nd Battalion even managed to produce a platoon sergeant who compared quite well to his regular colleagues.

As a commanding officer wrote in the *Gazette* at the time, those conscripts had made a great contribution to the services, as had the services to them. Those he commanded were ready to acknowledge that they were better citizens for their time in uniform.[15] Those who served in regiments such as the Green Howards were, of course, lucky. Rather less appreciative were many who had spent a couple of humdrum years in some administrative unit at home: life there was rather less of a challenge.

★★★

For the time being there remained in a corner of Richmond Barracks a small Regimental Headquarters, together with the museum. With its staff of two retired officers, this RHQ has done much to strengthen the Regiment's links, at first with the North Riding and later, after the local government reorganization of 1972, with the new counties of North Yorkshire and Cleveland. Through the benevolent fund, to which ninety-nine per cent of serving soldiers subscribe, it grants help to any old soldier or widow who may be in need. As well as looking after the museum, of which more later, this small staff edits the *Green Howards' Gazette* which at the time of writing approaches its centenary. It takes a keen interest in those local units of the Army Cadet Force that wear the Regimental badge. And, of course, it runs the Green Howards' Association. One of the strongest in the country for a Regiment of its size, with nearly 9,000 members and thirty-one branches, scattered between Newcastle and Dorset, this association is a live concern. Each year, as near to Alma Day as possible, some 500 old comrades are drawn to the weekend reunion, held in the old days at Richmond and later at Queen Elizabeth Barracks, Strensall. Its culminating event is always the church parade at which the veterans, medals sparkling, march proudly behind their Branch standards past the Colonel of their Regiment. It is a sight that can stir the heart even of unmilitary civilians watching from the roadside.

Among the myriad other tasks of Regimental Headquarters is the organization of 'Richmond Sunday'. Every spring a rather smaller gathering marches past the Regimental War Memorial at the top of Frenchgate before attending a service in Saint Mary's Church and taking tea afterwards in the town hall. In Saint Mary's is the Regimental Chapel, the stark simplicity of its Yorkshire stone and its oak carved by the famous Yorkshire 'Mouseman', Robert Thompson of Kilburn,[16] set off by the green and scarlet of the old Colours of the Regiment.

★★★

It took only a short time for the Yorkshire Brigade to become a close-knit community of North Yorkshiremen, both regular and Territorial. But further convulsions lay ahead. For the Green Howards the worst effects fell upon the Territorials.

When the National Serviceman finished his time with a regular unit, he went on to complete three and a half further years reserve service in the TA. Reformed in 1946, the 4th Battalion was still based in the north of the Riding, but the 5th Battalion had become gunners, their title 631 (Green Howards) Light Regiment, RA; however, they still wore the Green Howard collar badge and carried the old 5th Battalion Colours. Recruiting at first was slow, and the two units were kept in being only by the devotion of a dedicated and enthusiastic hard core. Then, at the height of the Korean War in 1951, the government gave this handful of volunteers the task, not just of training the reservist National Servicemen, but also the 'Z' Reserve which soldiers discharged at the war's end had been obliged to join. It was a bold decision. That summer the 4th Battalion carried to camp the seeds of serious trouble: 800 of these reservists, who had never taken their 'Z' obligation in any way seriously, were tacked on to a mere 100 volunteers and 300 of the reluctant National Servicemen. But all went well thanks to the common sense of all concerned. It was a valuable experience for the few volunteer officers and senior NCOs who had to command and train numbers exceeding their wildest dreams. They did this ably. During 50th Division's manoeuvres on Salisbury Plain in 1953, a party of foreign military attachés could be persuaded only with difficulty that they were not watching a brigade attack by regular troops. The changes in strategy already described brought about the next upheaval among the Territorials. With any major war postulated to last no more than a week or so before nuclear weapons brought it all to an end, no need could be foreseen for large numbers of reservists. So it was that the role of the Territorial Army became ever more vague. The first result was the amalgamation in 1961 of the 4th Battalion and the Light Regiment, RA, into a 4th/5th Battalion, a happy arrangement but one that did not last.

A further and far more severe blow fell in 1966 when the government announced the disbandment of the Territorial Army and its replacement the following year by the much smaller but better equipped Territorial and Army Volunteer Reserve. In this the 4th/5th Battalion disappeared and the Yorkshire Volunteers were created, their 1st Battalion including a single 'B' (Green Howards) Company based on Middlesbrough. An angry public reaction caused the government to give the matter more thought. A further unit, the Green Howards Territorials, its companies at Middlesbrough, Scarborough and Guisborough, then came into being. In contrast to the quite sophisticated equipment of the Yorkshire Volunteers, these new Territorials were only lightly armed, their primary role being the support of the civil authorities in a national emergency.[17] The Green Howards Territorials existed for only two years, but from its cadre grew in 1971 'D' (Green Howards) Company of 2nd Yorkshire Volunteers, based on Scarborough and Guisborough.

It should be said that, despite the welcome given to those Boer War Volunteers on their return to Richmond in 1901, it took not just the First World War but the Second as well for Territorials to be properly accepted as members of the Regiment. Only in 1925 could Militia and Territorial officers join the Dinner Club;[18] few ever did so. Not until the 1960s, when the annual officers' dinner moved from London to Yorkshire, were wartime, National Service and Territorial officers seen there in appreciable numbers. These days such officers hold prominent positions on the Regimental Council and its associated committees.

The cuts mentioned above fell not only on the Territorials but the regular army also. Still more units were disbanded, including the York and Lancaster Regiment, another sad loss to Yorkshire. Too small now to be viable, these infantry brigades were in 1967 regrouped into 'divisions' (so further complicating the nomenclature of the British Army), the Yorkshire Brigade amalgamating with the North Irish and Lancashire Brigades into the King's Division, its headquarters in York and its Depot at Strensall.

The Green Howards had survived yet another upheaval, one of but five historic British line regiments never as yet amalgamated.

Above: *A 1st Battalion patrol operating in a flooded Belfast street in 1970.*

Below: *The Ardoyne, Belfast, to which the 1st Battalion moved in 1971 in the early days of the 'troubles'.*

12. PROFESSIONALS AGAIN, 1968-91

Life has a habit of coming up with the unexpected – and especially for soldiers. Just as few experts would have predicted that British armed forces would be fighting in the Gulf in 1991, one year after the Berlin Wall had been pulled down, even fewer would have thought that the army would still be committed to internal security duties in the United Kingdom more than twenty years after the first troops were deployed on the streets of Northern Ireland. Fewer still would have anticipated that the Regimental Band would have been chosen to represent the 1st Green Howards in the reconquest of Kuwait – not in their role of musicians but that of medical orderlies.

Despite the many tasks given to the Regiment, Northern Ireland has been the dominating feature of Green Howards' soldiering during the most recent period of the Regiment's history. From their first operational tour in 1970 – a year after the first infantrymen from our fellow Yorkshire regiment, the Prince of Wales's Own, were sent into the Province to come to the aid of the exhausted and overstretched Londonderry police – until 1991, the Green Howards have completed nine battalion tours of duty. In 1992 they were due to return there once again.

★★★

In 1967 the British Army had withdrawn from the barren rocks of Aden, once a coaling station for British ships en route to India and the east. In that same year the Government issued its Defence White Paper, confronting the economic reality that the country was a second-class power struggling precariously between the superpowers, and declared that what was left of Britain's imperial and global roles would be liquidated by 1971 and that our strategy and therefore our forces should concentrate on Europe.

So it was that the Green Howards found themselves in the late 1960s stationed in Colchester as part of the newly created Army Strategic Reserve, having returned from a short tour in Hong Kong and Malaysia in 1966 where they were within a few days of deploying on operations in Borneo when the 'confrontation' (as it was diplomatically described) with President Sukarno of Indonesia ended and Borneo's independence preserved. It is curious that the Regiment in the last twenty-five years has frequently been earmarked for action, and sometimes been within twenty-four hours' notice of going, only to find itself not needed: Borneo in 1966, Aden in 1967, the Falklands in 1982 and the Gulf in 1990. Northern Ireland has been a noteworthy exception.

Nevertheless the jungle skills learnt in the Far East were not wasted as, in 1968, the Green Howards provided the garrison of one company group in British Honduras, what is now Belize, for two nine-month tours without their families. This was one of our last colonies, often neglected by London which historically has looked east, but still requiring British forces to protect it from invasion by its neighbour Guatemala. It was a popular posting, especially for young platoon commanders and their men who were able to disappear into the jungle for independent training, away from the eye of the company commander or more importantly the sergeant-major. In addition, the splendours of the Caribbean coral-reef coast were there for everyone to enjoy. The carefree Belizean people were easy to like and the two companies responded by immersing themselves in many community projects, from dispensing first aid in inaccessible jungle villages to helping build a new school on the coast. To this day, after twenty-five years, the Green Howards are remembered there with warmth. How different was life for the modern soldier, compared to his 18th and 19th-century forebears who served in the Caribbean and died from disease by the hundred.

In November 1969 the Battalion was posted to the British Army of the Rhine in Minden to become mechanized infantry as part of 7 Armoured Brigade – the Desert Rats. But before they had the opportunity to get familiar with their new role, they were despatched in June 1970 to the Falls Road area of Belfast for their initial Northern Ireland tour, first to protect the Catholic population and then to face its antagonism. It was a bewildering and often chaotic time, as well as a raw introduction to the complexities of Ulster politics.

Since the 2nd Battalion's time in Tipperary and Limerick in 1922, there had been a cyclical wave of violence in Ireland, first in the South following Partition, coupled with a Protestant backlash in the North in which some 230 people were killed and 1,000 injured. The Irish Republican Army was established as the military arm of Sinn Fein following the 1918 elections. From the thirties to the fifties it tried to mount terrorist campaigns which never succeeded because of lack of public support. The next 'Troubles' had begun in October 1968 after a series of protest marches organized by the Northern Ireland Civil Rights Association, largely on behalf of the Catholic minority. The Catholics had many genuine grievances about housing, job opportunities and political gerrymandering. In Londonderry in August 1969 these spilled over into violence which then spread in October to Belfast. It was during the serious rioting of early July 1970, in the grimy narrow streets of the Lower Falls, similar in many ways to the poorer parts of Middlesbrough, that the Green Howards were obliged to exchange their riot shields for rifles when they came under sniper fire. Indeed during the night of 3 July a curfew was imposed to avoid further bloodshed, a curious sight in the United Kingdom. In that month the tea stopped flowing in the Falls, and by the end of the year the honeymoon period between the British Army and the Catholic community had ended. The IRA campaign against the British Army had started. The Green Howards spent the rest of their two-month tour keeping the peace between the Protestant and Catholic ghettos in various parts of Belfast, only lightened by some ironic rescuing of inhabitants of the Falls during the floods in August, a humanitarian act not always fully appreciated by the women who only weeks before had been hurling abuse and bricks at the soldiers. It was important to maintain a sense of humour and one corporal, under attack, caught the mood when he gave the order – as if on the rifle range – 'Tell the company commander that we will paste up, move back to 300 yards and give two rings when we are ready!'[1]

★★★

The year 1970 was also marked by the introduction of the 'military salary'. 'Apart from the ending of conscription, it was probably the most significant alteration to military life since the war.'[2] Previously soldiers had been paid a relatively low salary, supplemented by free board and lodgings and by marriage allowance to officers over twenty-five years of age and to soldiers over twenty-one. It was difficult to encourage young men to 'join the professionals', the recruiting cry of the day, when they were in effect paid pocket money and the true value of their wages was hidden. The new scheme was aimed to equate with equivalent civilian trades; at first it worked well but it took a major pay rise in 1978, after low morale and high wastage, to restore full comparability. Nevertheless, despite criticism, the new system was generally fair and pay became less of an issue of concern. One consequence of abolishing marriage allowance, which previously had discouraged men from marrying young, was that the Green Howards, like other units, gained a far greater number of families to look after. This happened at a time when accompanied overseas postings were less frequent, and was a further inducement to the army to become more home based and sedentary – a tendency which grew with home and car ownership and with not just officers educating their children privately, but some sergeants and corporals also. In many ways this period of strategic and social change for the army in the late sixties was an unsettling time at regimental level, and the Green Howards lost a significant number of promising young officers who left to seek new careers in other professions.

★★★

A further administrative change made in the year following the introduction of the military salary was the granting of permission to the Green Howards to revert to their Regimental cap-badge, a much welcomed event. The white rose badge of the Yorkshire Brigade was, however, retained for use by the Yorkshire Volunteers, whose affiliated Middlesbrough and Scarborough companies continued to keep their links with the Green Howards.

But 1971 is better remembered for 1st Battalion's next tour in Ulster, the most violent a British battalion had as yet experienced. Since the previous year the incidence of terrorism had grown alarmingly, with bomb explosions and shootings a daily occurrence in Belfast. In a situation where the police had lost control in many Catholic areas, where mobs roamed freely, and witnesses were intimidated, the government reluctantly introduced internment as a desperate measure to restore public order. Centred largely on the strongly republican Ardoyne area of Belfast, the Battalion faced an unprecedented level of street violence, gun battles and terrorist sniping. The worst followed the introduction of internment on the night of 9 August 1971, the brunt being borne by two companies. This was before the days of well prepared and sophisticated pre-tour training and the Battalion learnt the hard way with the murder of Private Hatton on 8 August, one week after arriving in Northern Ireland, the first Green Howard to lose his life in this campaign. Although four more young soldiers lost their lives in two months and some forty were wounded, the Battalion, under the strong leadership of Lieutenant-Colonel Ronnie Eccles, gave as good as it got, accounting for a considerable number of terrorists and dominating the area, both tactically and psychologically. Further success came with the arrest of the city's two leading terrorists, Martin Meehan and 'Dutch' Doherty, on the afternoon of 9 November. It was an exhausting tour but one in which the Regiment gained a fierce reputation for courage, skill and resoluteness, which few regiments have subsequently bettered. Among the gallantry awards for the four-month tour was a DSO for the commanding officer, an MC for a platoon commander, an MBE for the RSM and – most significantly – three MMs for section commanders. It was to be above all a corporals' war.

Satisfied with what they had accomplished, at the end of the tour the Green Howards marched through the streets of Middlesbrough, as they had done so often in the past, where they received a tremendous reception by the crowds.

Such was the problem of finding enough troops for Northern Ireland that the 1st Battalion was back in Belfast for a further tour less than a year later, from October 1972 to February 1973, this time in the equally testing area of Andersonstown. The clash between the Parachute Regiment and the Catholic inhabitants of the Bogside area of Londonderry which had taken

Left: Lieutenant-Colonel R. Eccles, in command of the 1st Battalion, with Lieutenant-General Sir Ian Freeland, GOC Northern Ireland, in Belfast in 1970. Colonel Eccles later received the DSO, a rare award for Northern Ireland.

place on 30 January 1972 – known as 'Bloody Sunday' – brought serious political consequences. The bombing campaign reached its peak that year, and the IRA set up 'no-go' areas in parts of Londonderry and Belfast. Such local anarchy in parts of the United Kingdom could not be allowed to continue and on 31 July 1972 Operation 'Motorman' was mounted to retake these 'no-go' areas, eleven battalions being deployed in Belfast alone. The combination of 'Bloody Sunday' and 'Motorman' was to entrench still further the hatred of the Catholic community for the British Army.

By the time the Green Howards were in action that autumn, the security forces had the terrorists far better contained. Assisted by improved intelligence, the infantrymen were developing the patience and guile to outwit the gunman and bomber and so became master of this cat-and-mouse war; they were also learning the age-old lesson that it was a battle for the hearts and minds of the local community, one often waged in vain.

Despite the alertness and operational successes, another soldier was fatally wounded by Protestant gunmen in East Belfast on 7 February 1973. It was not just the IRA who were shooting at the army; this unhappy incident happened during a sharp gun battle, when the Green Howards were ordered to send a company to East Belfast to reinforce another regiment during serious rioting. Here a platoon commader, later to command 1st Battalion, won the Military Cross for leading his men from doorway to doorway under accurate sniper fire, and a sergeant from another platoon was awarded the Military Medal. More than a dozen terrorists were accounted for that hectic day, one an ex-Irish Guardsman, a further example of the tragedy of Ulster.

★★★

On 25 July 1973 the Colonel-in-Chief, His Majesty King Olav V of Norway, opened the new regimental museum in Trinity Church, Richmond. This was the happy culmination of many years' hard work by

Right: *The Colonel-in-Chief and the Colonel of the Regiment, Major-General D. S. Gordon, at the opening of the Regimental Museum at Richmond in July 1973.*

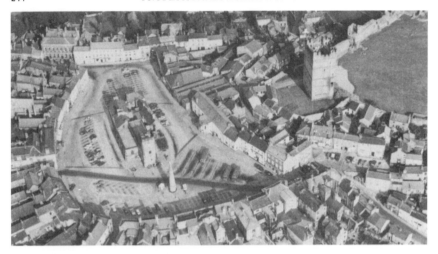

Above: Air photograph of Richmond. The Castle can be seen on the right. Regimental Headquarters and the Museum are housed in the redundant, 14th century Church of the Holy Trinity, seen in the centre of the Market Place. (Yorkshire Life)

Left: Painting of Brigadier T. F. J. Collins, by Carlos Sancha, which hangs in the Regimental Museum for which he was largely responsible.

the Colonel of the Regiment, Major-General Desmond Gordon, and the Regimental Secretary, Colonel Jonathan Forbes; the generosity of many subscribers from the county and beyond helped to give the museum the independence it may need when further government cuts occur. As Major M.L. Ferrar, the Regiment's principal historian, wrote: 'About thirty years ago [1894] the First Battalion began to collect a small collection of war medals, and very soon got the opportunity of buying the Victoria Cross and Legion of Honour of Private John Lyons at an auction in London.'[3] As the years went by a collection was built up of medals in the hope that:

> When future generations of Green Howards look on these medals they may picture to themselves the intrepid advance of the Regiment up the heights of the Alma, and the dreary winter of 1854, in the frozen trenches before

Sevastopol; the march up the rugged slopes of the Black Mountain to the Muchai Peak, 10,200 feet above sea level; they may think of the two sweltering summers on the Nile, when the crepe was never off the drums; the picketting of the Dwatoi heights and the bivouacs on the sterile uplands of the Tirah; of the rain of bullets at Paardeberg, and the long and thirsty marches from the Modder to the Crocodile river; of the weary days and nights in the water-logged trenches in Flanders and the incessant shelling and costly attacks on the various long-drawn battle fronts.'[4]

Major Ferrar himself established the groundwork for the museum, following which Brigadier Tommy Collins devoted his creative gifts and outstanding energy to its development into the archetype of a first-class county regimental museum; it reflected not only the Regiment's history, but also its close ties to the local community in which it lives and from which it draws its life blood, something unique to the British Army. Indeed in 1975 the museum was placed second in the nationwide Museum of the Year Award as a 'fine example of totally independent enterprise, determination and fund-raising'.

Despite the effort devoted to operations in Northern Ireland 1st Battalion managed to complete in the summer of 1972 an armoured battle group exercise in Canada on the prairies of Alberta; there units from the British Army of the Rhine could user the large tracks of land so necessary for modern armoured warfare training, and which were not available in Germany. It also carried out a full training season in Germany the following year. One of the strengths of the British infantry is its ability to adapt itself to quite different roles with flexibility and commitment; although operational skills are inevitably lost to a certain degree when so doing, the compensation had lain in its ability to cope with the unexpected, a quality that has often stood it in good stead. That, and the enormous benefit of realistic all arms training on the Canadian prairies, was to be of immeasurable value when in due course it was faced with armoured warfare in the Middle East.

★★★

Home in January 1974 to Chester, the only garrison unit in the city and thus a popular posting, it was not long before its fourth tour in Ulster was under way. By this time, however, the army had introduced an excellent 'package' of pre-tour training, specifically designed to match the area to which a unit was going; during this and later tours, such specialized training was to save many lives and increase the army's effectiveness. Often, however, it also lengthened the time a battalion was in 'baulk', and thus unavailable for its primary NATO assigned role.

This fifth tour saw the headquarters and one company of 1st Green Howards at Portadown but the rest of the companies scattered around the rural areas south-west of Belfast under command of other units. Although law and order had much improved and the RUC were carrying out a more traditional police role, political tensions were still running high. The experiment in Direct Rule from Westminster had ended in December 1973 after nine months, and the Sunningdale Agreement produced a power-sharing formula involving all sections of the community in a new Executive to replace the old Assembly. One month later, on 23 January 1974, the Loyalists walked out of Stormont and, in a deepening political crisis, the Ulster Workers' Council called a general strike on 15 May. Its effectiveness proved that without the support of the majority the power-sharing Executive had little chance of healing the divisions. Direct Rule was re-imposed in June.

For the Green Howards the summer tour of 1974 was best remembered for the efforts to keep Protestant tempers cool across barricades of farm-tractors, and for reassuring frightened Catholics in such flash-points as Obin Street in Portadown; the sight of soldiers manning petrol pumps added some welcome amusement. Not all the so-called 'patch' was free of terrorist activity however. A part of the Battalion, under command of another regiment, faced an elusive IRA enemy in the fields of east Tyrone where the techniques of countering skillfully laid bombs and landmines were being mastered. Tragedy struck again in Dungannon on 5 June when a soldier was shot dead on patrol. A local person, who witnessed the murder, sent a letter to the Colonel:

> What a nice boy he must have been. We saw four soldiers on the street, one very tall and grown up. The others looked much younger and all looked happy together as they usually did. As they passed the corner, shots rang out and one boy fell badly wounded. Two of his pals, still in the line of fire ran to him and

Left: The 1st Battalion Band playing in York Minster in 1974, shortly after its instruments and other property had been destroyed by an IRA bomb.

dragged him to safety. One knelt down and tried to bandage his poor head. The tall one stood up and cried out in grief and agony like a child who had lost a loved one. And still crying he shouted a warning to someone who had ventured onto the street. I would pray to have friends like these.[5]

It was the Regimental Band's turn, on tour that summer in Yorkshire, to come up against the terrorists' bombing tactics of hitting 'soft targets' in mainland UK. On the night of 10 June 1974 a bomb exploded in Strensall Barracks, totally destroying the Band's intruments and uniforms; only good luck prevented many human casualties. Not to be daunted, five days later they were playing on Richmond Sunday with borrowed instruments. The real battle was to follow over the next few years to get the government to agree compensation. This involved lobbying and debates in both Houses of Lords and Commons. In the end strong all-party pressure in the Lords forced the government to change its mind and pay up.

South Armagh was the operational area for 1st Battalion's next Northern Ireland tour from April to August 1975. The press had called this deceptively peaceful rolling patchwork of farms, thorny hedges and low stone walls 'bandit country'. Even before the present 'Troubles' erupted, it had always a well-earned reputation as a lawless border area, dominated by independently minded Catholic farmers, famous for smuggling, theft and illicit poteen stills. As a local saying goes: 'Between Carrickfergus and Crossmaglen, there are more rogues than honest men.' Little perhaps changes in Ireland. Back in the 1840s the Regiment had been employed tracking down such stills during the Whiteboy disturbances in Rathkeale; now such work was left to the police, while the rifle companies concentrated on the terrorists. South Armagh, with a long and meandering border giving easy escape into the Republic of Ireland, provides perfect conditions for guerrilla warfare. It was there that the IRA was to gain so many initial successes against the security forces. On the other hand, it was also a countryside where the infantryman could pit his patrolling skills of reconnaissance and ambush against a tough and experienced adversary.

During a long and hot summer the Battalion soldiered hard, often with little sound intelligence, mounting intricate and carefully designed operations to outwit and catch the terrorist. As Lieutenant-Colonel Peter Inge, then in command, and at the time of writing General Sir Peter Inge and Colonel of the Regiment, commented:

> The thing that impressed me most about the tour was the enthusiastic and constructive approach by the junior leaders and soldiers themselves. They were remarkably confident when they arrived but at the same time were prepared to learn and even in the most testing situations they never allowed things to get on top of them – I genuinely found them an inspiration.[6]

The training before the start of this rural tour had been of special importance. The Commanding Officer's drive and imagination in directing this training was not only a major factor in ensuring the success of the tour, but his methods were also to be taken into use by future battalions preparing for South Armagh.

A further Military Cross and Military Medal were to be awarded to members of the Regiment. But it paid dearly. On 17 July a company commander, Major Peter Willis, and three other soldiers attached for an operation at Cortreasla Bridge, near Crossmaglen, died when an enormous land-mine exploded. The sole consolation was that an observant corporal, later to be RSM, saw a man running away who was later arrested, convicted and sentenced for all four murders. It says much for the spirit and continuity of a family regiment that both Peter Willis's son and his nephew later became officers in the Green Howards.

***

The following year the Battalion managed to release companies for training in Cyprus and Malta before it was recalled to Ulster for a month's emergency tour in April and May 1976. It was back to County Down and Armagh. This was the first tour since 1970 when lives were not to be lost, so it was tragic indeed that only hours after getting back to Chester from Liverpool docks three officers were killed in a car accident a few hundred yards from the barrack gates.

Another overseas posting was due. The Battalion had been warned for Singapore, but because the government decided to withdraw the garrison, in 1976 the Green Howards were instead sent for two years to Berlin, a city that since the end of the Second World War had been divided into four occupied zones, garrisoned by the four victorious powers, Britain, France, the Soviet Union and the USA. After the seemingly endless run of Northern

Left: Major Peter Willis who was killed by an IRA bomb in Northern Ireland in 1975.

Above: Members of the 1st Battalion parade with Russian soldiers
at Glienicke Bridge, Berlin, on 24 August 1976, when the recently
discovered body of a Russian soldier, killed in 1945, was handed
over to his compatriots.

Below: The 1977 Queen's Birthday Parade in Berlin. Immediately
in front of the Colours is Lieutenant-Colonel L. G. James,
commanding the 1st Battalion.

Ireland tours during the previous six years, Berlin was a welcome respite,
especially for the families. Guards and duties, never very popular with
soldiers, took up much time but the proximity of East Germany border
guards and the nearby Soviet forces kept everyone on edge. For many it was
to be their only close contact with the Russian Army. During this 1978 tour
the Green Howards also enjoyed the honour of taking part in the Queen's
Birthday Parade in front of Her Majesty inside the impressive Maifeld
Stadium.

★★★

In September 1978 the Green Howards moved to Aldergrove, on the
outskirts of Belfast and near the airport, for an eighteen-month tour; it was

their first residential tour in Northern Ireland with their families. Operations were mostly in Belfast and often under command of other regiments, never a happy arrangement for the Colonel who was always losing his troops as 'renta-companies', as they were termed. Although by 1977 there had been a marked reduction in the general level of violence in the Province, and especially of deaths among the security forces, callous and cruel inter-sectarian violence continued. However, it was possible to start the gradual handing back of overall responsibility for the direction of the security campaign to the Royal Ulster Constabulary; for Green Howard foot patrols this often meant eight soldiers protecting one police officer to enable him to carry out his traditional law and order duties, such as delivering court summonses.

In March 1980 the Battalion returned to the same Catterick barracks from which it had sallied forth to war in 1939, a move marked on 9 May by a visit from King Olav, its Colonel-in-Chief. Hardly had the men and their families settled in than the Battalion flew off to Kenya for a six weeks' exercise. On its return it found itself guarding one of Her Majesty's prisons in County Durham, during a prison officers' strike. Then, in May 1981, the Green Howards for the first time donned the blue United Nations beret to carry out a six-month tour in Cyprus, keeping the peace between Turks and Greeks. On return, they took part in the final training of the Falklands task force. As ever variety was the spice of life.

January 1983 found the Battalion back in Germany, this time in Osnabruck as part of an armoured brigade in Rhine Army. For the next four years all ranks mastered their armoured personnel carriers and the intricacies of combined arms tactics on a modern battlefield. Infantry weaponry and equipment had grown increasingly sophisticated in recent years with the introduction of laser range-finders and hand-held computers for the mortar platoon; Milan guided missiles for the anti-tank platoon; and thermal imaging night-sights that turned night into day. (A hard-wearing but comfortable pair of infantry boots was still awaited.) Even the 7.62mm self-loading rifle was about to be replaced by the 5.56mm SA80 rifle with its magnified optical sight and automatic fire capability. In many ways it was a long way from the days of their fathers in Malaya or grandfathers in Normandy.

Although the infantryman was no longer the poor relation in terms of modern technology, it did not obviate the hard fact that he was still required to carry out his age-old task of closing with the enemy and so, more than any other arm, exposing himself to acute danger. This required the inculcation of an offensive spirit which was possible to achieve on training in Germany and in Canada. It could, however, work against the need to use minimum force and show acute political sensitivity when on internal security duties in Northern Ireland. The army – and more importantly the British public – asks a great deal from their young soldiers, and especially the junior NCOs, who often carry the responsibility of making crucial decisions in split seconds in Belfast, Londonderry or South Armagh. This was demonstrated during the 1st Battalion's next four-month tour in Ulster, this time in West

Belfast from June to November 1985. Although such tours from Germany took the soldiers' eyes off their main NATO role, it provided a change, especially for the young single soldiers, who were glad to get away from the routine of garrison life in Germany to the realities of anti-terrorist operations. It certainly sharpened up a battalion.

★★★

Although the original hope had been that the Battalion would be posted to Yorkshire so as to celebrate the Regiment's tercentenary in 1988, in the event it found itself back in Ulster in January 1987 on a two and a quarter year tour in Londonderry, again accompanied by its families. With the border so close, it was only too easy to feel one was on the north-west frontier of the United Kingdom. It was in Londonderry that the army had first been deployed in 1969, and names such as the Bogside and Creggan had become household words over the years. Two of the first things to strike the soldiers – apart from the welcoming brick – was that the city was much smaller than expected. Furthermore the place had its own very special character, distinct from the rest of Northern Ireland, with both sides of the communal divide fiercely loyal to it in their different – and differing – ways.

There was little time for reflection. The tour began with a high level of terrorist activity that required a firm military presence in close cooperation with the RUC. This intense pace of operations was to continue throughout the tour. Most of the other two-year battalions sent their companies off to reinforce other units, so the Green Howards were fortunate in having their own 'patch' and hence a sense of commitment and responsibility to success in Londonderry. The soldiers were able to get to know the streets, the local people, the moods of the city and thus become highly effective in developing a sixth sense that kept them one step ahead of the terrorist. During this time the IRA had obtained, mainly from Libya, large quantities of weapons and explosives, in particular the odourless Czech Semtex. This, coupled with the IRA's increasing sophistication and care in planning and mounting attacks against security forces, meant that the Battalion's tactics had to adapt constantly to the changing threat, whether it was mortar bombs, booby-traps or anti-armour grenades.

The intense patrolling, although often unpopular with some sections of the local community, had evident success in deterring and containing the IRA; it forced them to concentrate on 'softer' targets. As the Battalion was living 'over the shop' with married quarters very close to the city, the families started coming under attack in November and December 1988 when bombs were planted in hijacked cars outside their houses, causing damage but fortunately no serious casualties. Nevertheless, this cowardly act failed to intimidate the families or to drive any of them back to England. (The last time Green Howard families had been threatened in this way was by EOKA in Cyprus in 1955, but there they were not directly attacked.) Off-duty restrictions had already been tight in the Province, in particular after Corporal Metcalfe was murdered in a booby-trapped van in Lisburn in June 1988, having taking part in a charity fun-run. Although necessary, these

Left: A soldier of the 1st Battalion on patrol in Londonderry city centre in 1988.

Above: Lieutenant-Colonel F. R. Dannatt, commanding the 1st Battalion, confers with RSM Warriner on exercise in 1990.

Right: Men of 'A' Company being deployed by Chinook helicopter.

Above right: Colour Sergeant Ferguson driving an All Terrain Mobile Platform.

restrictions were such that one wag likened life to that on an oil rig, with periodic leave in mainland UK as the only freedom available.

It had been the 1st Battalion's ninth tour in Northern Ireland. Although very few men indeed had experienced them all, many had been to and fro many times and wore the single green and purple ribbon of the General Service Medal with pride, a decoration that tells the observer little about its wearer's length of service in that troubled province. It is sad indeed that the British have so seldom been generous in recognizing their army's service. In the words of Lord Lytton, the Victorian writer: 'What is a ribbon worth to a soldier? Everything! Glory is priceless.'[7]

★★★

In March 1989 a tired Battalion returned to Catterick to join the newly raised 24 Airmobile Brigade, where it faced the challenge of converting to its fresh role as an airborne anti-tank battalion, as well as preparing for the forthcoming presentation of new Colours by the Colonel-in-Chief and the tercentenary celebrations in June.

On 19 November of the previous year a small group of officers, together with the RSM, had dined in Dunster Castle, where three hundred years

before to the day Colonel Francis Luttrell had raised his Regiment. Among the company was Colonel Walter Luttrell, a direct descendant of its founder and the then head of the family. Sadly King Olav, who was to have honoured the officers with his presence, fell ill with flu in London and was unable to attend. The same day a group of 1st Battalion soldiers set off from Dunster to Exeter to re-enact the march that Luttrell's Regiment completed to join Prince William of Orange. But it was not until the next summer that celebrations could begin in earnest and over the weekend of 24/25 June 1989 King Olav, on his last visit to the Regiment, presented new Colours to the First Battalion and took the salute as his Regiment, followed by well over 400 Old Comrades, marched proudly past the royal dais. Afterwards 2,400 people, from King and Field Marshal to private soldier, sat down to lunch together in one massive marquee; that evening balls and dances took place in all the messes. The following morning, on a clear but windy Sunday, King Olav joined the Regiment for a church service in the grounds of Richmond Castle, with the newly consecrated Colours draped over the drums, in a fashion little changed over the centuries. It was a moving end to a weekend of pageantry and kinship.

★★★

There was no time for self-congratulation. After a week of freedom marches round our counties of Cleveland and North Yorkshire, created from parts of the North and West Ridings of Yorkshire and of County Durham after the local government reorganization of 1972, it was off to Germany for a major exercise in the airmobile role. Hardly had this finished than one company, strengthened by support weapons and the recce platoon, departed to the Falkland Islands and South Georgia for a six-month tour of duty from October 1989 to March 1990. While there, they were able to pay their respects to the grave of Captain John Hamilton outside Port Howard. He had died when serving with the SAS in the 1982 Falklands War, showing 'supreme courage and sense of duty by his conscious decision to sacrifice himself on behalf of his signaller'.[8] His gallantry was marked by a posthumous Military Cross; it could well have been an even higher award.

While based in Yorkshire and building on the publicity of the tercentenary, it was an ideal opportunity to concentrate on recruiting. Following the disbandment of the 4/5th Green Howards in 1967, and the formation of the Yorkshire Volunteers, it had been difficult to keep the cap-badge in the public eye when the Green Howards were out of England. County ties were in need of strengthening and renewing. The success of these recruiting efforts was seen in the fact that the Battalion was able to raise a fourth rifle company. Indeed in 1990 the Green Howards was the best recruited infantry regiment in the British Army. Moreover its soldiers were local men: in January 1991 only thirteen per cent of them were other than Yorkshiremen or Clevelanders. Unlike most of their predecessors, forty-nine of a total of 436 soldiers questioned owned their own homes and seventy-six were educated up to 'A' or 'O' level standard.[9]

But in certain ways the Regiment was the victim of its own success. At a time when many other regiments were seriously undermanned, the Green Howards were ordered to send a company to Northern Ireland in 1991 to help another regiment, which was understrength, to meet its commitments. But this still allowed time for the Battalion to exercise in Kenya in February and March 1991. During this period also the Battalion reinforced its recent tradition of travelling and adventure training, by sending expeditions as far afield as Borneo, Nepal, the USA, Africa and Cyprus.

On 17 January 1991 His Majesty King Olav died at the age of 87, having been Colonel-in-Chief for 32 years. General Sir Peter Inge, the Colonel of the Regiment, the Corps of Drums and a street lining party were present at his funeral. His strong affection for and personal interest in the Regiment were to be sorely missed. As His Majesty wrote in 1983 on the 25th anniversary of his appointment:

> Above all, I want to assure you of my enduring sense of attachment to the Regiment started by my Grandmother, Queen Alexandra, over a hundred years ago. The relationship between my family and the Green Howards is indeed unique, extending beyond this personal connection to the wider context of friendship between Britain and Norway.'[10]

13. A NEW WORLD ORDER: 1991-2001

With the fall of the Berlin Wall in 1989, burying European communism in the rubble, the last decade of the 20th century was to witness a thoroughly unstable and unsettled world after the relative certainties of the Cold War. One of the first fracture lines to open had been in the Middle East with the invasion of Kuwait by Saddam Hussein in August 1990, leading to a remarkable international coalition operation to drive the Iraqis out the following February. Hard on its heels, a second fracture line appeared in the Balkans in 1992. Against this backdrop, alliances that would have been unimaginable in the Cold War began to reshape international cooperation, replacing superpower confrontation with the hope of a New World Order.

The Green Howards were to play their part, but it was not until the middle of the period that the Regiment was deployed to the Balkans; earlier it was left to the Regimental Band to steal the glory, by providing medical assistants to the field ambulances supporting the allied troops into Kuwait in February 1991. It was the ongoing suffering of Northern Ireland that was first to claim the 1st Battalion's attention.

Throughout 1991 and all of 1992 the 1st Battalion, as already mentioned, continued to be in such a well-manned position, eighty-seven overstrength at its peak, that it was able to provide a fourth rifle company on six-month rotation to Londonderry to another regiment. At the same time demands for the services of the other three rifle companies continued. In 1991 the Battalion deployed to South Armagh on a short emergency reinforcement tour of a month, and then in 1992, now under the command of Lieutenant Colonel Nick Houghton, it deployed once again, initially to East Tyrone in June that year and then to North Belfast until the end of January 1993. This heavy commitment did not stop the companies going abroad to Cyprus, France, Oman and Kenya and later the whole Battalion to West Canada on overseas training, a welcome change for the soldiers but an increasing strain on the families. It was a theme that was to be thrown into even sharper relief as the decade gathered pace in an Army growing smaller, increasingly undermanned and over-committed. British Governments of both persuasions found it difficult to resist international pressures to use its highly regarded armed forces to "do something" and to "box above its weight" in the New World Order. One of the strong fibres that held regiments together through these busy days was their family spirit, epitomized by the Laidlaw family with three serving brothers in the Sergeants' Mess. It was also the Regiment's ability to recruit so successfully that had again saved the Green Howards in July 1991 from sharing the fate of so many other old and famous regiments, which had been disbanded or amalgamated as a result of a drastic reduction in the size of the armed forces following the end of the Cold War.

In 1922 HM King Harald V of Norway succeeded his father as Colonel-in-Chief, further forging the royal family and Norwegian link with the Regiment. Straightaway he showed his interest in his new role, dining with a number of officers at the Royal Hospital Chelsea on 11 November 1992, and then visiting Dunster Castle and the 1st Battalion the following September.

The Army at this time decided to strengthen the links between its regular and territorial infantry battalions, part of this process being to revert to the old Territorial Army battalion names and numbers. In our case elements of the Yorkshire Volunteers changed their title to the 4th/5th Battalion The Green Howards on 25 April 1993 under the command of Lieutenant Colonel Wade Tovey, so that all Green Howards, both regular and TA, once again wore the same cap badge. This was a welcome move for those pressing for greater integration for the TA, but it was a sad moment for those who had built up a strong Yorkshire Volunteer ethos and identity over the past 25 years. But the TA soldier, as adaptable and resilient as ever, quickly adjusted to this new identity. For many in the 4th/5th Battalion, shooting success was of more immediate importance, crowned by Captain John Alexander's award of the Queen's Medal at Bisley in July 1994. But further change for the TA was just round the corner.

The following year, after an exercise in Dhofar in Oman, the 1st Battalion left Catterick in the summer for Osnabruck in Germany, its previous home from 1983 to 1987, to become an armoured infantry battalion equipped with the new and potent Warrior fighting vehicles. Before it left, as part of an Army-wide review of military music to raise musical standards and cut the number of bands, the Regimental Band held its last parade in Northallerton and was then disbanded. The newly arrived commanding officer, Lieutenant Colonel Andrew Farquhar, had the painful task of ordering the Band under Bandmaster Searle to "March off Parade" for the final time, witnessed by many a moist eye. The 1st Battalion Band had provided military music to the Regiment since 1774, stirring the heart and stiffening the sinews of countless soldiers down the generations. Shortly after the Battalion had arrived in Germany, it deployed to East Tyrone in Northern Ireland from December 1994 to June 1995, so delaying its conversion to its armoured infantry role until it returned. There was still time beforehand in July 1994 to provide a Royal Guard of Honour in Edinburgh for the Colonel-in-Chief's State Visit, as the Regiment had done for his father at the Royal Palace of Holyrood House in September 1962.

At the end of 1994 Field Marshal Peter Inge, who by this time had risen to be Chief of the General Staff and then Chief of the Defence Staff, handed over after twelve highly influential years as Colonel of the Regiment to Brigadier Richard Dannatt, now commanding 4th Armoured Brigade in which the 1st Battalion was serving. At his dining-out from the 1st

Above: The Colonel of the Regiment, General Sir Peter Inge, and Brigadier Maurice Atherton show the silver statuette of Princess Alexandra, presenting Colours to the 1st Battalion in 1875, to her great grandson HM King Harald V of Norway and Queen Sonja before dinner at the Royal Hospital Chelsea on 11 November 1992.

Above: Warrior armoured vehicles patrolling north of Mostar in central Bosnia during the hard winter of 1996/97.

Left: A (King Harald) Company, as part of a large NATO force, entering Kosovo in June 1999.

ove: In early 1995 the Regiment
inted an expedition to the world's oldest
deepest lake, Lake Baikal in Siberia, to
ry out a scientific survey of the
angered Nerpa seals. Cpl Kidger is
sing the frozen lake on an all-terrain
icle.

ow: The 2000/1 Regimental Biathlon and Nordic Ski team
h Lt Col Patrick Roberts. LCpl LS Jackson (rear right) was
cted for British Winter Olympic Team training.

Battalion the Field Marshal, when asked by Drum Major Daniels for permission for the Corps of Drums to fall out, replied "No" and the party continued late into the night.

Besides its hectic operational and training commitments, the Regiment still found time for sport and adventurous training – indeed it made this a theme of its recruiting message. While Captain Tim Rodber was playing rugby for England and the British Lions, other members were making their reputations in cricket, nordic skiing and boxing, winning Infantry and Army cups. Corporal Jackson, a talented biathlon skier, was quickly identified as a future Olympiad. In addition, expeditions were taking place in parts of the world as distant as Lake Baikal in Russia, South Georgia on the edge of the Antarctic and the Himalayas of

Above: *Before undertaking an operational tour in Bosnia in 1996, the 1st Battalion spent four weeks armoured training on the prairies of Alberta in Canada.*

Pakistan, all generously supported by the wisely invested Regimental funds.

In the summer of 1996 there was a unique meeting of the 1st and 4th/5th Battalions on excercise in Germany, as the regular Battalion hosted their territorial counterparts for their annual camp. The 1st Battalion then spent the remainder of the summer on the hot and dusty Canadian prairies of Alberta, in demanding armoured battlegroup training with some of their TA comrades working alongside them. They earned praise from all parties for the speed at which they had successfully mastered the art of manoeuvre tactics.

That summer the Regiment launched the Friends of the Green Howards Museum with an eighty-year commemorative exhibition. Galvanized by Colonel Geoffrey Powell, the Friends have grown to a battalion-size membership under Brigadier Bill Marchant Smith, Richard Leake and Major David Nicholson's energetic leadership, with an acclaimed newsletter edited by Major Roger Chapman that has twice won national awards. The Friends' most recent achievement has been to establish a stone memorial at Contalmaison in July 2000 to one of the Somme VCs, Second Lieutenant Donald Bell, the only English professional footballer to win his country's highest gallantry award.

The Regiment was one of only two in the British Army to land two assault battalions on D Day, when CSM Stan Hollis stormed across an orchard in Crépon in his second action to win the only VC that day. In recognition of this, in October 1996 a bronze statue of a Green Howard was unveiled in Crépon by King Harald, to commemorate those Green Howards who gave their lives in Normandy and in the Second World War. The imposing statue by James Butler is deeply moving in portraying an exhausted but resolute infantryman resting on a broken wall and reflecting on that historic day's events. A fighting regiment draws great strength and pride from the deeds of its forebears; their example sets high standards to maintain. One of those who had been most generously associated with the Crépon Memorial, Sir Ernest Harrison, provided a surprise postscript to the project. As a great admirer of heroes, Sir Ernest revealed that he was the owner of not only the Hollis VC medal set but also of Private Henry Tandey's medals – Tandey, as the holder of the VC, DCM and MM, was the most highly decorated private soldier who survived the First World War. At a reception in the Tower of London in 1997, hosted by Field Marshal

Right: *The Green Howards Memorial at Crépon in Normandy, sculpted by James Butler, was unveiled by the Colonel-in-Chief on 26 October 1996.*

The Lord Inge as Constable, Sir Ernest presented both medal sets to the Regiment, an immensely generous gift. They are now displayed in the refurbished medal room of the Regimental Museum in Richmond, appropriately named the Harrison Gallery.

In the last week of October, just six week's after taking over command, Lieutenant Colonel Lamont Kirkland took the Battalion and its sixty-three Warrior vehicles to Bosnia. They were to be part of the NATO Implementation Force (IFOR) in the Gornji Vakuf area. Known as the "Anvil", a Croat-Muslim Bosnian sector flanked on three sides by the Serbian Republic, the NATO troops' job there was to monitor compliance of the Dayton Peace Accord. The people of the Balkans have through the centuries been subject to either war or inter-communal murders. Since the former Yugoslavia disintegrated with the death of President Tito in 1980, there had been constant unrest; various regional leaders, most recently Slobodan Milosevic, had used racial distrust as a means of consolidating power. In 1992 Bosnia declared independence from Serb-dominated Yugoslavia; this in turn led to the Bosnian civil war. Repeated episodes of violence, military occupation and ethnic cleansing not seen in Europe since the Second World War followed. After unsuccessful United Nations intervention, followed by a more robust American and NATO political response, the Dayton Peace Accord, already mentioned, was signed in 1995. Thus ended three years of savagery, with Bosnia now separated into two autonomous regions: the Muslim Croatian Federation in the centre of the country, with the Serbian Republic around it. IFOR's task was to implement this fragile peace. The first Brtish troops to form the force were those of 4th Armoured Brigade commanded by the new Colonel of the Regiment, Brigadier Richard Dannatt.

Bosnia in winter is a less than friendly place with the temperatures dropping to minus twenty-five degrees, making driving in armoured vehicles along frozen mud tracks in the mountains a hazardous risk. Nevertheless the Battalion set about establishing its presence through a rigorous patrol programme to ensure freedom of movement, to reassure the population and to make sure the warring factions kept their weapons under lock and key. The presence of the Warrior vehicles was critical to this. As their Colonel remarked: "The former warring factions had to be made aware of what they would have to fight if they took us on in combat. At the same time, we tried to bring the various groups together and create a working entente. It was gratifying at last to see them together, talking and planning and enjoying relations for the first time in five years. It was important to bring quick and instant relief to the area. Hence we spent a great deal of time on reconstruction projects – rebuilding a school, creating a children's playground, mending a health centre and constructing a bridge. The bridge, later to be called the Bridge of Peace, was to bring together the Muslim and Croat communities who had been divided when the bridge was destroyed in the early part of the war. Part of the Dayton Agreement was to return the country to a multi-ethnic society. This perhaps was our most complex and sensitive operation. But finally people

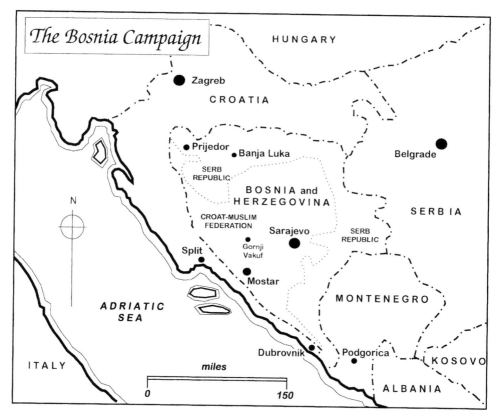

The Bosnia Campaign

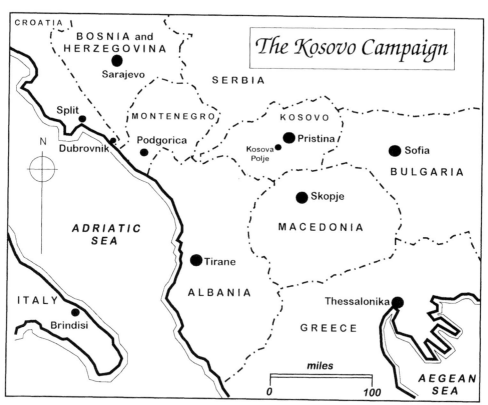

The Kosovo Campaign

did return to their homes whilst we were there – indeed we saw the roots of peace taking hold and beginning to spread."[1]

The British Army has won a reputation for having the versatility, professionalism, physical and moral robustness and sense of humanity to be the most sought-after peacekeepers and peacemakers in the international community. No UN or NATO operation ever seems to get off the ground now without regiments like the Green Howards as part of the force. This was all at a time when the increasing pressures of political correctness, risk aversion and litigation were starting to have an insidious and demoralizing effect on the warrior culture of the British Army. Afterwards the more prosaic garrison life back in Osnabruck was a welcome respite and the Battalion was able, for the rest of 1997 and all of 1998, to concentrate on its armoured infantry role with training in Canada, Germany and Poland.

In June 1998 a memorial was unveiled in the Regimental Chapel in Richmond by Field Marshal Sir Nigel Bagnall to commemorate those who had died on active service since 1945; of the twenty-nine names listed on the plaque, nineteen were killed in Malaya, many of them national servicemen. As a young Green Howard officer, Field Marshal Bagnall had won two Military Crosses in Malaya with the 1st Battalion. In the turbulent years that followed the Second World War, it is significant that there is only one year when a British soldier has not been killed since 1945 and that was in 1968, the year before troops were deployed to Northern Ireland.

Lieutenant Colonel Patrick Roberts took over command in December 1998. As a cavalry officer, it was a natural move for him to command an armoured infantry regiment in the modern era of close all arms cooperation. There was also a nice continuity in that his father, Colonel Michael Roberts, had commanded both the 1st and 2nd Battalions in the 1950s. As 1999 began, the 1st Battalion once again prepared to deploy on operations in West Belfast. However, events in the Balkans took an unexpected turn as Serbia's President Milosevic began applying violent pressure on the Albanian population of Kosovo. The international community resolved to stop him and NATO began operations to protect the Albanians of Kosovo. Hostilities had begun in Kosovo in 1998 where ethnic Albanians had been fighting for an independent state. Then the Serbs, who believe Kosovo to be the cradle of the Orthodox Christian Church and centre of their culture, launched a tide of racial cleansing to keep the province part of Serbia. Hundreds of thousands of Albanians fled across the country's borders into Albania, Montenegro and Macedonia to escape the conflict. As these horrors in the Balkans escalated, it looked as if the whole of the 1st Battalion would be deployed to Macedonia (and from there into Kosovo), but it was not to be. In February A (King Harald) Company was ordered to stand by to join the 4th Armoured Brigade Group already in the Balkans. The remainder of the Battalion, much frustrated by being split up, was then reinforced by a company from the Black Watch and continued to prepare for Northern Ireland. Such was the state of overstretch in the Army that units were robbing Peter to pay Paul so as to be properly manned for operations. At the end of March 1999 NATO then began air strikes against selected Serb targets in an effort to force President Milosevic to the negotiating table.

For A Company, under the command of Major Simon Fovargue, there began an intense period of preparation and training in Osnabruck; at any moment they might have to move to Macedonia to join the rapidly increasing NATO ground forces preparing for an invasion of Kosovo. On 21 May, with no improvement in the political situation, the order came to deploy the company's 130 men and eighteen Warrior vehicles to an area south of Skopje in Macedonia to join the Irish Guards Battle Group. On arrival the company began a concentrated period of acclimatization and live fire training in the heat and dust of northern Macedonia. The uncertainty as to whether or not the NATO force would have to make a forced entry into Kosovo remained. After seventy-two days of non-stop NATO bombing, President Milosevic backed down and, on 4 June 1999, signed a peace agreement. The hope was that Kosovo Force (KFOR) could now enter the country unopposed and implement the UN Security Council Resolution 1244, signed just days before.

So it was that at 11 am on Saturday 12 June 1999 A Company moved north to cross the Macedonian border, passing en-route tented refugee camps full of Kosovar Albanians who ran out cheering and shouting

Below: Major General FR Dannatt, the Colonel of the Regiment and Commander British Forces, wishes Major SG Fovargue good fortune before he leads A (King Harald) Company into Kosovo in June 1999. Major Fovargue was later awarded the MBE for his leadership.

"NATO! NATO!" and throwing flowers at the armoured vehicles, whilst children were held aloft carrying placards proclaiming their gratitude. As soon as the troops crossed the border in the vanguard of a long armoured column, the first signs of war damage and massacre were there for all to see. The whole area was deserted, strewn with the evidence of a rapid exodus; abandoned cars, suitcases, clothes and even children's toys lay in the road. Burnt-out houses carried Serb slogans as a reminder of their activities. Their move was laboriously slow along the heavily congested route supporting the advance. As the weather closed in and night fell, they were forced to stop just ten kilometres south of Pristina, the capital and their objective, so as to allow the Serb forces, as part of the peace agreement, to begin their withdrawal back into Serbia. During the night an eerie procession of Serb armoured convoys trundled past with their occupants making three-fingered Serbian victory signs and waving their Kalashnikov rifles defiantly in the air; despite the protracted air campaign the Serbian Army remained largely intact and withdrew in good order. Two days later the company was ordered to secure Kosovo Polje, some fifteen kilometres to the south-west of Pristina. Known as the "Field of Blackbirds", it was where the Serb army was defeated by the Ottoman Turks in the 14th century and hence central to Serb nationalism; the place where Milosevic rose to power in the 1980s, it is one of the most politically sensitive areas in the region. As the company moved off to their task, they were accompanied by the BBC war correspondent Kate Adie, well known to the Regiment from Ulster, who reported to British TV viewers on "these humorous yet determined Yorkshiremen".

Throughout June to October the soldiers continued their roller-coaster ride of uncertainty and unpredictability. As Simon Fovargue commented, "We had trained for the worst case that might have involved heavy fighting in a forced entry into Kosovo, but now relied on residual Northern Ireland experience and a great deal of initiative from every single soldier. Demonstrations, intimidation, property destruction and looting were widespread. Operations varied from identifying and marking massed graves to the protection of life and property. It was a rude introduction for those young soldiers who had not seen dead bodies before. The soldiers coped admirably with the demands, stresses and strains placed on them."[2]

In mid-September 1999, as the company was preparing for its handover to the Norwegian Telemark Battalion, a two-week period of severe disruption began, culminating with a grenade attack by Kosovars on 28 September, leaving three Serbs dead and over forty civilian casualties. The Serbs had lost confidence in KFOR's protection and barricades were thrown up round their enclaves. Finally, the Brigade Commander lost patience with the Serbs' intransigence and ordered the company to mount an operation to remove all barricades by force. On 5 October A Company Group – of five British platoons and some sappers, a Norwegian platoon, a section of Canadian recce vehicles and a platoon of Italian Carabinieri – demolished the barricades and secured the area in fifty-three minutes. It was the first planned use of force by KFOR against the civil community in Kosovo. Athough negotiations continued with the Albanians, the task of persuading the Serbs and NATO to talk to each other continued to be an uphill struggle. In the last week of A Company's tour, the Serbs finally agreed to meet, but only to say 'thank you and farewell' to the Green Howards.

On leaving Kosovo, the Colonel of the Regiment, now Major General Richard Dannatt, who was the Commander British Forces there, paid the following tribute. 'For the Company it had been an exhausting and exhilarating experience from first to last. From the uncertainty of the waiting period in Macedonia, through the advance into Kosovo, on to Pristina with the enthusiasm of the welcoming crowds, and then the deployment to the sensitive area of Kosovo Polje with the explosive mix of Albanian, Serb and Roma populations. Throughout, A (King Harald) Company was superb, earning justifiable praise from all who came into contact with it.'[3]

At the same time the rest of the battalion moved to West Belfast in May, to undertake a six-month tour of duty as part of 39 Infantry Brigade, under command of the now Brigadier Nick Houghton. The situation was very different from previous tours, as a 'ceasefire' was in operation as part of the Peace Process following the historic Good Friday Agreement. Much of the Battalion's work was boring and monotonous, yet vital. Not only was the summer's marching season kept relatively peaceful, but various attempts by dissident sectarian groups to use violence to disrupt the peace process were thwarted by a network of observation posts in Belfast in

Right: The Secretary of State for Northern Ireland and MP for Redcar, Dr Mo Mowlam, in characteristically informal mood with two of her constituents, LCpl Cooper and Sgt Ray, on a visit to B Company in Belfast on 29 May 1999.

support of the RUC and covert agencies. The end of a busy six months presented some curious but telling statistics: the Battalion had spent 67,000 man-hours watching over Belfast; some 10,000 photographs had been taken; over 1,000 hours of video footage had been shot, all of which led directly to the RUC making more than 200 arrests. In addition the Battalion was responsible for the recovery of £1.2 million worth of stolen goods. This painstaking and methodical approach was in sharp contrast to the hectic and embryonic days of the early 1970s. As Lieutenant Colonel Patrick Roberts remarked: 'In 1999 we deployed to two operational theatres simultaneously and acquitted ourselves with distinction in both, whilst retaining a stable rear base in Osnabruck: some achievement indeed.'[4]

In England that spring and summer the Territorial army was suffering another metamorphosis following the Government's recent Strategic Defence Review. It was decided to reduce its size by a third, with the major impact on the Infantry. Lieutenant Colonel Duncan Hopkins was the Commanding Officer who received the bad news that 4th/5th Battalion was to be disbanded; the compensation was that the two companies in Middlesbrough and Scarborough were to be retained in the new Tyne Tees Regiment and keep their Green Howard cap badge. The old 50 (Northumbrian) Division sign of the double T was also to be revived. It was a pragmatic British compromise. At the heart of this review, based at the Headquarters of the Infantry, was a Green Howard, Colonel Andrew Farquhar. The Colonel of the Regiment took the final salute at the disbandment parade of the briefly revived 4th/5th Battalion at Wathgill Camp, Catterick Garrison, on Sunday 21 March 1999.

In April the following year the 1st Battalion left Germany after almost six years and moved to Warminster in Wiltshire as the Land Warfare Training Centre Battle Group, in earlier days known simply as the Demonstration Battalion. It was good to be back home and, although the soldiers worked hard during the week to provide demonstration troops and act as enemy on armoured exercises on Salisbury Plain, they were able to enjoy a more stable and predictable home life and get away north to Yorkshire and Teesside at weekends.

Also the return to England enabled the 1st Battalion to devote much needed time and effort to recruiting: Germany had not been popular with many soldiers and as a result the Battalion was in the unfamiliar position of being understrength by some eighty soldiers. However, through coordinated and sustained effort, including drawing on Regimental funds, by late 2001 the situation had improved with a twin-track campaign of better retention and more recruits joining.

Meanwhile the devastating outbreak of foot and mouth disease across many farms in the country in 2001 forced the Government to call upon the Army's logistic and organizational skills to bring the crisis under control. In late March, while the commanding officer was in an audience with the Colonel-in-Chief in Oslo, the 1st Battalion deployed initially under the Second in Command, Major Julian Panton, to assist the civil authorities in the Midlands. As the Battalion arrived in Worcester, taking on responsibility for

Above: Cpl Oakes and LCpl Skeen about to light a pyre to dispose of cattle carcasses infected by Foot and Mouth Disease in the West Midlands in 2001.

the counties of Shropshire, Herefordshire and Worcestershire, it became apparent at the outset that the situation was confused and the local authorities were in danger of losing control. Although the details of foot and mouth disease were totally unfamiliar to most soldiers, sound military common sense and training were quickly applied; teams led by NCOs were sent out to the farms to gather information, an operations room was established and the Battalion took responsibility for the disposal of the carcasses. Throughout this bizarre and sad task the soldiers were greeted with courtesy by the local farmers and many were invited into the farms for meals and to relax. The Provost Sergeant, Sergeant Gent, in charge of finding combustible materials for funeral pyres, regularly ordered in excess of £10,000 of materials daily. The Battalion left Worcester in late May after a farewell lunch with their civilian colleagues, many of whom had no previous contact with the Army but had learnt to respect their adaptability and professionalism. The Green Howards' legacy was to deal with over 300 infested premises and the disposal of some 300,000 dead animals. As a final gesture the Commanding Officer was made an honorary life member of the National Farmers Union!

On a happier note on 23 April 2001, St George's Day, the Regiment was proud to hear that Field Marshal The Lord Inge of Richmond was to be invested as a Knight of the Garter by Her Majesty Queen Elizabeth II. It says something for this small county Regiment that in the last seven years it has produced one Field Marshal, one future Lieutenant General, one future Major General and five Brigadiers.[5]

Left: Field Marshal Lord Inge processing as a Knight of the Garter at Windsor.

The New World Order was to display many bizarre characteristics during this turbulent decade, perhaps typified by the Moster bank raid led by the Colonel of the Regiment a week before St George's Day, on 17 April 2001, on the head office of Herzegovacka Bank in Southern Bosnia. Deployed as Deputy Commander (Operations) of the Stabilization Force (SFOR) in Bosnia, Richard Dannatt and the SFOR leadership found themselves challenged by a breakaway Bosnian Croat leadership that wished to form its own mini-state in the south of the country, Herzegovina. Self-rule independence was the public rallying cry, but protection of racketeering and corruption was the real reason. The only way to gain evidence was to enter the head office of the Herzegovacka Bank and obtain proof about its money-laundering activities. With the city of Mostar asleep and secured by French and German battlegroups, the British battlegroup raided the bank at 0200 hours, removed five truckloads of documents and computers, blew the safes securing over three and a half million Deutchmarks of currency and returned to barracks for an early breakfast - an extraordinary episode, and all entirely legal!

The infamous terrorist attacks on the World Trade Centre in New York and the Pentagon in Washington on 11 September 2001 will be etched on everybody's memory, shaking nations and individuals into a realism that the last decade, with its hope of a New World Order, had ended brutally in an apocalyptic day of destruction not seen since the Second World War. As President George W Bush declared 'the first war of the twenty-first century', politicians and military commanders throughout the civilized world search for a solution to the new global threat of transnational terrorism. But what is certain is that the British nation and its tried and tested Army will continue to need the services of such county regiments as the Green Howards to defend our freedoms and values.

EPILOGUE

In the summer of 1991, when the final pages of the first edition of this book were being written, the collapse of the Soviet Union and the outcome of the Gulf War dominated world events. Now, ten years later, with the conclusion of such an eventful ten years, as the 1st Battalion prepares for a further tour in Ulster in Spring 2002, it is worth reflecting that, despite the recent clutches of the Balkans embracing the British Army and the subsequent war in Afghanistan, it is Northern Ireland that remains a major focus of the Regiment's operational attention. The problem, as always in a democracy, is to defeat terrorism without resorting to unacceptably draconian measures. Indeed there is a prospect that this restraint and patience over more than a quarter of a century have now borne fruit, giving Northern Ireland some hope for a more peaceful future. The Green Howards are proud to have played their part. Some ninety awards for service in this tragic corner of the United Kingdom bear witness to this. Nine Green Howards have died while doing their duty there.

If it is asked how the Green Howards and other similar regiments have managed to maintain their high professional standards, fortitude and compassion in Northern Ireland, returning year after year until now to discover that scant political progress has been made, one answer may surely lie in the regimental system.

During more than 300 years the value of this regimental system has been incalculable. It is the envy of other armies in which the fighting man, lacking the stable background of a regimental home, moves frequently from one unit to another. It has, of course, its defects. A regiment is primarily an enlarged family, and a hierarchical one at that. Such families can be conservative in outlook and suspicious of outsiders. The regimental system can also involve the taxpayer in some slight extra expense, a fault seized upon by those who struggle to ensure the survival of balanced and effective armed forces within an ever-decreasing budget. The difficulty lies in attempting to compare the logic of a balance sheet with the priceless but intangible benefits the regimental system provides in sheer fighting efficiency. There is a need always to remember that:

> The Army is not like a limited liability company, to be reconstructed, remodelled, liquidated and refloated from week to week as the money market fluctuates. It is not an inanimate thing, like a house, to be pulled down and enlarged or structurally altered at the caprice of the owner, it is a

living thing. If it is bullied, it sulks; if it is unhappy it pines; if it is harried it
gets feverish; if it is sufficiently disturbed, it will wither and dwindle and
almost die, and when it comes to this last serious condition, it is only revived
by lots of time and lots of money.[1]

Winston Churchill said this in 1904. Changes to the Army's size and shape
are always needed, but those responsible for them, both military and
civilian, should bear these words in mind.

More than half a century ago another famous man, Field Marshal
Lord Wavell, that highly intelligent and understanding soldier, described
the outcome of all military training as the private soldier 'advancing of his
own free will in front to face the enemy'. What he said had held good as
much at Fontenoy and the Alma as it had done on the Somme. Despite the
complexity of the electronically controlled weapons to which the modern
Green Howard has adapted himself, Wavell's remark applies in exactly the
same way today. Any battle may well end at the bayonet's point. This
happened even among those technological complexities of the 1991 Gulf
War, as it had in the Falklands nine years before. In the end, success in
battle may lie in the infantryman advancing to his front in the face of
enemy fire or in defending his position when all hope of relief has been
extinguished – dying alone in a muddy ditch, unseen and but briefly
remembered. Far more battles are lost through the failure of the soldier to
fight than through poor generalship.

But sound morale is needed as well as proper training if soldiers are to
be persuaded to make the ultimate sacrifice. The British infantryman seldom
finds inspiration in a profound cause. Nor do appeals to patriotism or *la gloire*
stir him. Religious fanaticism he fails to understand. His strength lies in
simpler things: his confidence in his own abilities and his leaders; his sense of
duty; and, most important of all, the knowledge that his friends expect him
to behave in a certain way. A corporal of another line regiment waiting for the
ground war in the Gulf to start, said, 'Obviously people are wondering . . .
how they will react to combat. But I honestly think the fear of being shown
up in front of other people will overcome that. Nobody wants it to get home
that he lost his nerve or did nothing to help a mate who was hit.'[2]

Today, as seldom before, the majority of regular recruits for the Green
Howards have their homes in the industrial towns of Teesside, the fishing
ports and holiday resorts of the North Yorkshire coast, and the farmlands of
the inland moors and dales. Many are the sons, nephews or grandsons of
Green Howards; some take pride in a yet lengthier lineage. Tribal and tight-
knit they certainly are, moved by the fierce esprit-de-corps so evident when
they celebrated their tercentenary at Catterick in 1989. But those are the
qualities that make them into magnificent fighting soldiers. And as like is
attracted to like, it is those qualities that continue to attract recruits into the

Right: *His Majesty King Olav V, King of Norway, Colonel-in-Chief
1958-90, painted in 1968 by Edward Halliday. He took a very close
interest in his Regiment.*

Green Howards. And so as often has happened in the past when a battalion of the Green Howards has been shattered in action, semi-trained reinforcements, too often complete strangers to its background and codes, manage in some near inexplicable way to acquire strength from the handful of survivors, behaving in a few weeks as if they had spent a lifetime as Green Howards.

The core of the army's strength will always be rooted in such regiments as the Green Howards, its Battalions manned often by regular professional soldiers, sometimes by its part-time enthusiasts and so superbly well in the past by its conscripted National Servicemen.

NOTES

Chapter 1. Early Days: 1688-97 17-32

1. *GHG*, v 1, 1893, 8. 'The Raising of the Regiment' by John Parker (Major J.V.R. Parker) quotes Treasury Papers, v 13, no 45; HO Records; and Narcissus Luttrell, *Brief Relations of State* or *Diary of John Evelyn*, without giving detailed references.
2. Ibid.
3. *GHG*, v 75, 1961, 598, 'P.V. Investigates' by JMF (Colonel J.M. Forbes) discusses Lieutenant-Colonel P.V.V. Guy's examination of the Luttrell Papers and correspondence PVV-JMF.
4. Little, 109-10.
5. Traditional story.
6. Maxwell Lyte, v 2, 205-7; DNB.
7. DNB.
8. Neighbour, Part 1, 2-3.
9. Ibid., 3.
10. Story, v 1, 24-32.
11. Ferrar, 21-2.
12. Story, v 1, 103.
13. Ibid., 114.
14. Ibid., 27.
15. Macaulay, v 3, 681.
16. Story, v 2, 129-31.
17. Ferrar, 30-31.
18. GHG, v 11, 1903, 89.
19. Neighbour, Part 1, 3; Scouller, 69.
20. Ibid., 3-4.
21. Ferrar, 34.
22. Ibid., 8.
23. Fortescue, v 1, 354.
24. Ferrar, 8.
25. DNB.
26. Neighbour, Part 1, 1.
27. D'Auvergne 1693, 93.
28. *GHG*, v 12, 1904, 193-5, 'The Rawdon Papers'.
29. D'Auvergne 1694, 75.
30. Fortescue. v 1, 378.31. Ferrar, 39.
32. Ferrar, 34-5.
33. Childs, John, 'Lord Cutt's Letters, 1695'. Campden Miscellany XXX, Royal Historical Society, 1990.
34. Copy proclamation of 23 February 1699 in possession of author.

Chapter 2. The French Wars: 1697-1793 33-56

1. Neighbour, Part 4, 1-2.
2. Ibid.
3. Ibid., Part 1, 1.
4. Ferrar, 45. However, a subsistence account in Erle's papers suggests that the unit may have embarked for Cadiz only seven companies strong. If this is so, nothing is known of what happened to the other five companies.
5. Scouller, 208.
6. Ferrar, 46, quoting Burchett.
7. Neighbour, Part 5, 1.
8. Ibid., Part 6, 1-2. The letter was probably written in 1706.
9. Ibid., 1.
10. Ibid., Part 7; Cannon 8; Fortescue, v 1, 505, states that he never encountered the name of anyone below commissioned rank who fought in these wars.
11. Neighbour, Part 6, 1.
12. Ferrar, 46, gives the year as 1707. However on 13 April 1708 Erle received a letter from Marlborough mentioning that three extra battalions would be coming to Flanders, 'including your own'. (Neighbour Part 6, 1). Nor does Erle's appear on the Flanders 1707 order of battle; no order of battle for 1708 appears to exist. (Fortescue, v 1, 491 and 511)
13. DNB.
14. Fortescue, v 1, 507; Parker, 79.
15. Fortescue, v 1, 525; Ferrar, 50; *GHG* v 17, 1909, 98-100, *The Battle of Malplaquet*. Boyer, one of main contemporary accounts, does not mention the Regiment's presence at the battle.
16. Neighbour, Part 7, 2.

17. *GHG* v 76, 1968, 623-6, 'The Dress of the Green Howards in 1709' by W.J. Carman, FSA, FRHistS.

18. Norman, xxiii, the main authority on the subject, has written: The whole question of the award of battle honours abounds with anomalies. Petty skirmishes have been immortalized and many gallant fights have been left unrecorded. In some cases certain corps have been singled out for honour; others which have an equal share in the same day's doings have been denied the privilege of assuming the battle honour.

19. *GHG*, v 70, 1962, 229, 'Admissions Register Grove's Foot, Royal Hospital, Chelsea'.

20. Ferrar, *Bygone Days*, 92-3.

21. Several are reproduced in *GHG*, v 11 and 12, 1903 and 1904, 'Some Letters of Brigadier-General Richard Sutton'.

22. *GHG*, v 38, 1930, 195, 'Expedition to Vigo', quoting Dublin State Papers, by Major M.L. Ferrar.

23. Ibid., 196.

24. Ibid., v 45, 1937, 113, 'Expedition to Vigo 1719' by The Dragon.

25. Ferrar, 63.

26. *GHG*, v 13, 1905, 43-4, 'Service in Scotland': Extracts from the *Scots Covenant* 1739.

27. Ibid., 44 and 59.

28. Francis Grose, *Military Antiquities Respecting a History of the English Army*, v 2, G Hooper, 1786, 248. It should be noted that Captain C.R.B. Knight, *Historical Record of the Buffs*, Part 1, Medici, 1935, 130-1, quotes Ferrar as the source for that Regiment's name.

29. Some writers, including Eric and Andro Linklater, *The Black Watch*, Barrie & Jenkins, 1977, have given the Green Howards credit for having been especially prominent in the retreat. However, F.H. Skrine (the principal authority on the battle) in a letter written in 1907 (*GHG*, v 61, 1933, 194) could find no evidence for this, suggesting that there might have been confusion between the Regiment and the Buffs.

30. Ferrar, 76.

31. Fortescue, v 2, 154.

32. *GHG*, v 6, 1898, 68-70, 'Garrison Orders, Gibraltar'.

33. Ibid., v 10, 1902, 139, 'Memoir of Lieut-Col William Rickson' by JB.

34. Ferrar, 85.

35. *British Minor Expeditions compiled in the QMG's Department*, HMSO, 1884, 16-20; USI, v 43, 1899, 161-83 & 520-33, *The Siege and Capture of Belle Isle 1761*.

36. *GHG*, v 6, 84-7, 'Garrison Orders, Gibraltar'. Huttchison's name fails to find mention in Ferrar, *Officers of the Green Howards*.

37. Ibid., v 32, 1924, 190-1, Correspondence the American War of Independence.

38. McCrady, 333; Marion, 225.

39. It has often been said that the Regiment lost its mess silver at Monck's Corner. This is unlikely. Expert opinion at the National Army Museum believes that the acquisition and presentation of silver was not instituted until after 1790.

40. *GHG*, v 41, 1933, 139-40, 'One Hundred and Fifty Years Ago' by JB; DNB.

41. Nugent Papers, 6807-179, 19th Regiment Hospital Register, provides details of the variety of diseases that beset the Regiment.

42. Ferrar, 103-4.

43. *GHG*, v 83, 1975, 35. 'A Letter from General David Graeme' by J.M.F. (Colonel J.M. Forbes).

Chapter 3. The Kandyan Wars and Napoleon: 1793-1820 57-78

1. Fortescue, v 3, 519.

2. Ferrar, 104-5 gives departure date as 6 November. Reports by the brigade commander show that the Battalion was in camp before Menin on 21 September and that it embarked at Ostend on 13 October. WO1/167, 179, 219, 309.

3. WO1/167, 557, gives rank and file strength on 1 July 1794 as 527 fit for duty and seventy-five sick. Casualties may, therefore, have been in the nature of 40 per cent.

4. A full account of this little-known episode is in *GHG*, 21, 1913, 102-5. It is based upon Captain Robert Percival, *Cape of Good Hope*, 1804. An officer of the 19th Foot, he was also one of the principal authorities for the early history and topography of Ceylon. *See* Bibliography.

5. Beaver, *Services*, 437.

6. Percival, 394.

7. Although something is known of the areas the Regiment visited, nothing has come to light to indicate how it fared on these operations.

8. Cordiner, v 2, 262.

9. Percival, 82.

10. Ferrar, 111-3.

11. Cordiner. v 2, 188.

12. Anderson, *Wanderer in Ceylon*. This letter is contained in an Appendix which was included in only a small number of copies, the only one known by the author being in the Colombo Museum.
13. Anderson, *Poems*, 189.
14. Alexander, v 1, 337.
15. Marshall 106.
16. Beaver, *Services*, 710.
17. Johnston, 61.
18. Alexander, v 1, 164.
19. Peasley, 16-7, quoting Lockyer's Diary.
20. *GHG*, v 21, 1913, 155-6. 'Statement Relative to the Messing and Necessaries of the 19th Foot' signed by Lieutenant-Colonel William H. Rainsford who was second in command, a battle experienced ex-Guardsman commanding temporarily in the absence of Lieutenant-Colonel, on leave. An inspection report of the same month states that Rainsford had rendered 'the 19th Regiment one of the most respected in His Majesty's service (Ferrar 161). Stuart became a general and Rainsford a major-general.
21. Calladine, 43.
22. *GHG*, v 11, 1903, 90, 'Death Rate in Ceylon'.
23. Peasley, 22-3, quoting PRO WO 27/116.
24. Turton, 43-54.
25. Ibid, 67-8.
26. *GHG*, v 46, 1938, 225-6, '5th Bn History (1794-1807)' by A.J. Mackenzie.
27. Ibid., v 7, 1899, 196-7; v 78, 1970, 752-3, 'The Masham Independent Company, North Yorks Volunteers, 1795-1805', by Captain Tony Warrington.
28. Powell, *Kandyan Wars*, 198-9.
29. Calladine, 50.
30. Anderson, *Wanderer in Ceylon*, 121.
31. Powell, *Kandyan Wars*, 256.
32. Calladine, 77.

Chapter 4. From Napoleon to the Crimea: 1820-56 79-94

1. Calladine, 84.
2. Ibid., 101-2. The unfortunate Mrs McDonald was widowed the following year. Her husband, who had served in Travancore, went on half pay in October and died soon afterwards.
3. Ibid., 110.
4. *GHG*, v 21, 1913, 186-7, 'Field Marshal Lord Strathnairn's Memoirs: Service in Ireland with the 19th Foot'.
5. Calladine, 122.
6. Ibid., 188; Powell, *The Green Howards*, 50, is in error.

7. Ferrar, 190-2.
8. *GHG*, v 38, 1930, 105-6, 'An Episode In 1829'.
9. Calladine, 153-5.
10. Ibid., 168.
11. Ibid., 172.
12. Ibid., 196.
13. *GHG*, v 37, 1929, 71-2, Editorial.
14. Ferrar, 200-1.
15. *GHG*, v 33, 1925, 171-2, 'Some Notes on the Band and Drums of 1st Battalion The Green Howards'.
16. *GHG*, v 71, 1963, 56-7, 'Marches Presented to Lieut-General the Hon Sir Charles Howard by The Empress Maria Theresa Including Correspondence Between Col D'A Mander and Mr OW Neighbour of the British Museum'.
17. Francis Grose, *Military Antiquities Respecting a History of the English Army*, v 2, G Hooper, 1786, 248; *GHG* v 33, 1925, 189-92, 'Some Notes on the Band and Drums, 1st Battalion'; v 52, 1949, 47-9, 'A Brief History of the Band and Drums of 1st Bn The Green Howards' by Mr R. Lester, ARCM.
18. Usherwood, 4.
19. Calladine, 193.
20. Turton, 124-32.
21. *GHG*, v 4, 1896, 94-6, 'In The Crimea With The 19th Regt' by Margaret Kerwin.
22. Ferrar, 208.
23. *See* Note 21.
24. Russell, 55.
25. *GHG*, v 3, 1894, 'The Battle of The Alma' by John Kerwin.
26. Usherwood, 35.
27. Russell, 70.
28. Usherwood, 19.
29. Longford, Elizabeth, *Wellington: Pillar of State*, Weidenfeld & Nicholson, 1972, 11, quoting Wellington Despatches, 8 Aug 1815.
30. *GHG*, v 80, 1972, 1164, 'With The 19th Regiment in The Crimea'.
31. *GHG*, v 5, 1896, 89-92, 'The Battle of the Alma' by Major G. Lidwell.
32. *See* Note 25.
33. *See* Note 31.
34. Kinglake, v 2, 338.
35. Usherwood, 56-7.; W01/370/5.
36. Usherwood, 94.
37. *GHG*, v 10, 1902, 184-7, 'Christmas 1854' by Major G. Lidwell; Usherwood, 105.
39. Fortescue, v 13, 216.
40. *See* Note 21.
41. Ferrar, 247.

Chapter 5. Queen Victoria's Small Wars: 1856-1902 95-122

1. Kipling, Rudyard, 'Widow at Windsor', in *Collected Verse* Hodder & Stoughton, 1949 ed, 414.

2. Mott's book was not appreciated by all latter-day Green Howards. Nevertheless, when it appeared it was enthusiastically reviewed in *GHG*, v 6, 1898, 100-1; a copy was held in the Depot Officers' Mess and extracts were published in *GHG*, v 37, 1920, 50-1. The book contained many exaggerations but was accurate in its essentials.

3. Ferrar, 260; Mott, 109-19.

4. Mott, 101-2.

5. Ibid., 200. The pice was the smallest Indian copper coin.

6. *GHG*, v 79, 1971, 1001-2, 'The Grenadiers of Her Majesty's 19th Regiment' by Brigadier T.J.F. Collins. *See also GHG*, v 73, 1965, 255, 'An Old Photograph *c*. 1868' by Colonel J.M. Forbes.

7. For further information *see GHG* v 66, 1958, 112-3, 'The Green Howards' Hunt' by Captain I.R. Kibble.

8. The main regimental sources for this campaign are Ferrar, Neville and *Frontier and Overseas Expeditions*.

9. *Frontier and Overseas Expeditions*, 125.

10. Ferrar, 358-61.

11. *GHG*, v 20, 1912, 651. 'Regimental News in the Sixties' (extracts from the *Army & Navy Gazette*).

12. Mott, 215.

13. *GHG*, v 32, 1924, 171, 'Regimental Promotions' by Major M.L. Ferrar, 1871-1920; v 2, 1894. 59.

14. *GHG*, v 44, 1936, 21, 'Editorial'; 40. 'The Colours'; v 70, 1962, 302-4,'The Colours of the Regiment'; Ferrar, 279-81.

15. *GHG*, v 1, 1893, 31, 'Regimental Football Past and Present' by A. M. Handley. The Football Association was founded in 1863.

16. *GHG*, v 30, 198-201, 'Halifax, N.S. 1880-83' by Major J.T. Cotesworth; v 98, 1990, 34-7, 'The Green Howards in Halifax – Garrison Life' by Ian Hollaway; Ferrar, *Bygone Days*, 12.

17. The main regimental sources for the action at Ginnis are *GHG*, v 79, 1971, 1028-9; 'The Action at Ginnis' (copy of a letter by Major J.H. Eden); *GHG*, v 1, 1893, 125-6, 137-8, 'With the Guns at Ginnis' by Captain W.L. Mercer; *GHG*, v 36, 1928, 163, 'The Action at Ginnis' (attributed to Major M.L. Ferrar): *GHG*, v 42, 1935, 'Editoria'; Ferrar, 291-4.

18. *GHG*, v 45, 1937, 111-3, 'Battle Honours'.

19. Ferrar, 304.

20. *GHG*, v 1, 1893, 43-7, 'Among the Kachins' by Lieutenant H.C.W. Williams.

21. *GHG*, v 32, 1924, 105-6, 'Our Gazette' by the Editor.

22. The main regimental sources for the Tirah are *GHG*, v 5, 173-4; 'Letters From The Front' by Captain Holmes; *GHG* v 31, 1923, 165-7, 183-5, 199-202, vol 32, 1924, 13-5, 'A Company Officer in the Tirah War' by Major J.T. Cotesworth; v 6, 1898, 94-6, 'An Orderly Corporal's Despatches on the Tirah Expedition' by Lance-Corporal F. Jones; Fife, 5-30; Ferrar, Ch 17.

23. Fife, 26.

24. Ibid., 26-7.

25. *GHG* v 12, 1904, 'With The Mounted Infantry in Somaliland'.

26. *GHG* v 39, 1931, 55 (statement by R. S. M. Riordan)

27. *GHG*, v 58, 1950, 171-2, 'The Regimental Dinner'.

28. Ferrar, *Bygone Days*, 35-6.

29. Ferrar, *The Green Howards in South Africa*, is the main regimental source. Subsequently A.P. Hatton's 'Memories of the South African War' (an NCO's view) appeared in *GHG*, v 39, 1931, 229-34, v 40, 1932, 30-4, 49-54, 114-6, 127-131; other sources are *GHG* vol 21, 1933, 208-10, TWS, *The Mounted Infantry*; vol 38, 1930, 212-4, 'The Battle of Paardeberg' (letter from an officer); vol 12, 1904, 41-3, 'The Battle of Paardeberg' (Translated from the official German account).

30. Maurice, 124-4.

31. Ibid.

32. *See* Note 29. A.P. Hatton, v 40, 33.

33. Kipling, Rudyard, *Collected Verse, MI*, Hodder & Stoughton, 1949 ed, 463.

34. Ferrar, *The Green Howards in South Africa*, 81.

35. Turton is the source for the 4th Bn. He makes only passing mention of the 3rd.

36. *GHG* v 9, 1901, 70-5, 'Welcome to the Green Howard Company at Richmond' (from *Yorkshire Post*).

Chapter 6. The First World War and its Prelude: 1902-15 123-42

1. *GHG*, v 74, 1966, 272-5 & 313-6, 'Those Were the Days' by B.V. Rhodes, describes Regimental life in India in the early years of the century.

2. *GHG*, v 66, 1958, 290-1, 'Regimental Personalities' by General Sir Harold Franklyn. Further articles were suppressed.

3. Ferrar, 351-4. In speaking to the Battalion when he relinquished command, its colonel emphasized its success in training competitions, saying that he thought it more important to excel in this way than in sporting events.

4. *GHG*, v 15, 1908, 161.

5. *GHG*, v 6, 1904, 17; Ferrar, 353-4.

6. Fife, 60.

7. Ibid., 57-9; *The Times*, 14, 15, 16 April 1911.

8. Edmonds, *OH France* 1914-15, v 1, 10-11.

9. *GHG*, v 81, 1974, 12, 'The Menin Cross-Roads Machine-Guns' By JM Forbes.

10. Atkinson, *7th Division*, 46.

11. Edmonds, *OH France* 1914-15, v 2, 123-4.

12. *GHG*, v 44. 1936, 231-2, 'The 2nd Battalion War Picture' by Colonel H. Earle.

13. *GHG*, v 72, 1964, 92-4, 'An Unwilling, Uninvited and Unwelcome Guest of Kaiser Bill' by Tom Riordan.

14. Atkinson, *7th Division*, 74.

15. Wynne, 49.

16. Ibid., 50. A draft of the officer and 78 other ranks had also joined during the battle. Figures do not quite tally with those in Atkinson, 108.

17. *GHG*, v 74, 1966, 298, 'The Green Howards' Link with Bridlington' by J.R. M.

18. Edmonds, *OH France* 1914-15, v 3, 177.

19. Wylly, 123-5.

20. Wyrall 17-8.

21. *GHG*, v 73, 1965, 22-3, 'Territorials in Battle' by Edwin King.

22. Edmonds, *OH France*, 1914-5, v 3.

23. Wylly, 126-7.

24. Wyrall, 18.

25. Simkins, 34.

26. Ibid., 176, quoting J.G. Gordon letter of 22 July 1965 (IWM BBC/GW)

27. Wylly, 170.

28. Fife, 61-3.

29. *GHG*, v 24, 1916, 9, 'Tenth Service Battalion' by Revd. O.B. Parsons.

30. Colonel A.C.T. White. Letter to author.

31. Simkins, 24, quoting W.S. Hamer, *The British Army Civil-Military Relations, 1885-1905*, Clarendon Press, 1970, xi, 213-22.

32. *GHG*, v 22, 1915, 235, '12th (Service) Battalion, Teeside Pioneers'.

33. Colonel A.C.T. White. Letter to author.

34. *GHG* v 82, 1974, 33, 'Joining Up in 1914' by Major F.L. Allen.

35. *GHG* v 82, 1975, 35 'Experiences of the First World War' by S.N.S.

36. Edmonds, *OH France*, 1914-15, v 4, 313-4.

Chapter 7. First World War: 1915-18 143-64

1. Aspinall-Oglander, 235.

2. Quoted by Terraine, *First World War*, 60.

3. White. Letter to author.

4. Fife, 65.

5. White. Letter to author.

6. Ibid.

7. Ibid.

8. Fife, 97-102, elaborates the description of the fighting given in Wylly and *OH France* 1916, v 1, 363-41. The premature attack is discussed in *GHG*, v 97, 1989, 'The Mystery of Major Kent' by John Sydney.

9. *GHG*, v 66, 1958, 290-1, 'Regimental Personalities' by General Sir Harold Franklyn.

10. Fife, 103.

11. Terraine, *First World War*, 118.

12. White. Interview with author and *GHG*, v 74, 1966, 107-109, 'The Somme Battle 1916 and the Somme Commemoration 1966'.

13. Wylly, 229.

14. Fife, 76-7.

15. Read, Herbert, *Poems 1914-1934*, Faber, 1935, 39.

16. Wylly, 94-5.

17. *OH France* 1917, v 2, 209.

18. Wylly, 346.

19. Ibid., 147.

20. *OH France* 1917, v 2, 209.

21. Quoted by John Terraine, *The Road to Passchendaele*, Secker & Warburg 1977, 342.

22. Wylly, 270-1.

23. Read, 228-54. This essay *In Retreat* was first published as a pamphlet in 1929. It is one of the finest pieces of descriptive writing about war; all Green Howards should read it or have it read to them. However, an anonymous reviewer in *GHG* v 37, 1929, 225, was happily dismissive, suggesting that 'with the hint I have given him he will now give us something even more worthy of our Regiment and our Army, good in its way as this little book may be.'

24. Wyrall, 334.

25. See Note 8. Kent, now a lieutenant-colonel, was killed commanding 4th Battalion.

26. Wylly, 118.

Chapter 8. The Inter-War Years: 1918-39
165-82

1. As well as Wylly's account of the part played by 6th and 13th Battalions in this campaign, I have depended mainly upon Ironside and Leader. Leader mistakenly describes the 13th as the 9th Battalion, an error elsewhere repeated.
2. Wylly, 201-2.
3. The extended voyage is vividly described in *GHG*, v 81, 1974, 11-13, 'Adventures on a Troopship' by 'Corporal' (Mr. L Brayshaw).
4. Leader, 669-70.
5. *GHG*, v 28, 1920, 61-3, 'The Red Soldier' by Captain L. Marriage.
6. Ironside, 110-1.
7. Ironside, 112-3. Wylly, writing close to the events, ignores this well-known and often exaggerated incident. It is covered in detail by Lawrence James, *Mutiny: in the British and Commonwealth Forces, 1797, 1956*, Buchan & Enright, 1987.
8. James, op. cit., Note 7, 122.
9. Wylly, 211.
10. Ironside, 149.
11. Wylly, 13.
12. Lieutenant-Colonel H.B. Morkill, Tape 13.
13. GH Regt Archives, File 183; Morkill, ibid.
14. For this Frontier expedition, I have depended upon Wylly, Molesworth, Swinson and *GHG* v 27 & 28, 'With the 1st Battalion in Afghanistan' by Lieutenant-Colonel M.D. Carey.
15. Holt, 116.
16. Sources for the 1919-22 Irish troubles are *GHG* v 28, 29, 30, '2nd Battalion Notes'; *GHG* v 72, 1964, 22-3, 44-5, 65-6, 'I Remember' by 'Chirby' (Councillor W. Woodall); Lowe; scattered references in Townshend.
17. Townshend, 216-9.
18. Calladine, 134.
19. *GHG* v 28, 1920, 84.
20. Ibid., 99 & 147.
21. Ibid., v 72, 1964, 23.
22. Ibid., v 29, 1921, 59.
23. Ibid., v 72, 1964, 66.
24. Townshend, 57.
25. *GHG*, v 29, 1921, 40.
26. Ibid., 26.
27. The mode of operation of these columns is well described in Lowe.
28. *GHG*, v 29, 1921, 40-1; 'Reports in

Historical Record 2nd Bn 1858-1942'.
29. Townshend, 177.
30. *GHG*, v 29, 1921, 59.
31. *GHG*, v 30, 1922 65-6, Editorial; ibid, 109-11, 'The Late Lieutenant K.R. Henderson, MC'; Hampshire Regt Journal, Jan 24, 18-20.
32. GH Regt Archives, File 100, Letter from Colonel Hugh Levin.
33. *GHG*, v 32, 1924. 'Regimental Promotion'.
34. This can be confirmed from the pages of the *GHG* and the reminiscences to be heard in many of the tapes in GH Tapes.
35. *GHG*, v 28, 1920, 18-9, '2nd Battalion Notes', give details of its work in and around the Old City.
36. Sources for Shanghai 1927 are *GHG*, v 35, 1927, '1st Battalion Notes'; Gwynn Ch 8.
37. Sources for Palestine are *GHG* v 37, 1929, '2nd Battalion Notes'; Gwynn Ch 9.
38. *GHG*, v 39, 1931, 13-15, 35-7, 'Some Aspects of Shanghai'; 8, 'Acts of Gallantry – Citations'.
39. Sources for Frontier 1937-8 are *GHG* v 45-6, 1937-8, '2nd Battalion Notes'; v 71-2, 1963-4, 'Memories of Waziristan' by Harbinger (Colonel J.M. Forbes); v 72, 1964, 'Letters from Waziristan'; v 90, 1982, 25, 'The North-West Frontier on Active Service (1936-9)' by Steve Donelly; Major-General D.S. Gordon, Tape 10; Colonel Styles Tape 17.
40. Sources for Palestine 1938-9 are *GHG* v 66-7, '1st Battalion Notes'.
41. *GHG* v 70. 1962, 305-7, 'The Battle of Beit Furik' by 'GAFS' (Major G.A.F. Steede).

Chapter 9. Second World War: Norway to Tunisia, 1939-43 183-98

1. Barnett, 423.
2. Montgomery of Alamein, Field Marshal, *Memoirs*, Collins, 1956, 50.
3. The main source for the Second World War is Synge, a meticulously researched book, its contents thoroughly checked by participants in the various battles. Some of his notes and many of the eye-witness accounts he collected are held in GH Museum Archives 58.
4. As well as Synge Ch. 2, there is a separate folder of accounts of the Norwegian fighting in Museum Archives 58, most of them written on the journey home. Other sources are GH Oral History, Tape 10C

Norway by Major-General D.S. Gordon and *GHG* v 92, 1984, 12-13, 'Otta Revisited June 1984' by Colonel A.D. MacKenzie, an account of a battlefield tour.
5. Museum Archives, 58. Unsigned account.
6. Fraser, 41.
7. Synge, 27.
8. Ibid., 30.
9. Ibid., 45.
10. Ibid., 70.
11. Mander, 6.
12. For an account of the last hours of 5th Battalion see *GHG* v 91, 1983, 21, 'Captain C.G. Browning Journal of a POW', Part 1.
13. Synge 106, quoting Brigadier Desmond Young, *Rommel*.
14. Ibid., 142-3.
15. Clay, 114.
16. Synge, 184.
17. Nigel Hamilton, *Monty: Master of the Battlefield*, Hamish Hamilton, 1983, 218.
18. Decorations for gallantry were awarded with far greater selectivity and in fewer numbers between 1939 and 1945. An officer transferred from another regiment in the early twenties arrived with an MC, an OBE and three 'Mentions' all won while working as an Assistant Provost Marshal in the previous war. Both DSOs and MCs were frequently awarded in the First World War for good staff work at rear headquarters.

Chapter 10. Second World War: Sicily to the Elbe, 1943-5 199-218

1. Both quotations are from Synge 212. For this chapter also, Synge is the major source, together with Clay and Aris. Hull again fills the detail and provide atmosphere.
2. *GHG* v 75, 1967, 369-70, 'Hedley Verity's Last Tour' by A.L.S. (Brigadier Arnold Shaw).
3. Clay, 218.
4. Aris, 145.
5. Ibid., 172.
6. Ibid., 186.
7. Fraser, 279.
8. As well as Synge and Aris, this account of Anzio owes much to Raleigh Trevelyan's *The Fortress*. Trevelyan, a Rifle Brigade subaltern, had been posted to the 1st Battalion. Major Peter Howell, Tape 6, provides information by Brigadier John Scott as well as himself.
9. Aris, 222-3.
10. Trevelyan describes this in graphic detail,

but Howell does not remember the incident.
11. Mander, Chs 8-10.
12. Only recently RHQ has received a letter from a member of that brigade's staff in which his superb qualities are recalled.
13. Synge, 284.
14. Museum Archives, 58. Post-battle report by Lieutenant-Colonel Richardson, commanding 7th Battalion; Clay 228.
15. Adding to the sources already quoted, *GHG* v 82, 1974, 9-17, contains a feature on the D-Day landings that includes articles by Lieutenant-Colonel C. Macdonald Hull, Lieutenant-Colonel R.W.S. Hastings, Colonel R.H.E. Hudson, 'Timber' Wood, Colonel R.J.L. Jackson and Major Don Warrener. Additional material by Hastings is in *GHG* v 80, 1973, 'An Infantry Battalion in Battle – Two Actions' (reprinted from the British Army Review, v 1, 1949).
16. For many post-war years the British Army Staff College organized a battlefield tour of Normandy for its students. The team included generals as well as Lieutenant-Colonel Robin Hastings and C.S.M. Stanley Hollis. Hollis is said to have been the star of the show, telling his story in an extremely modest way and bringing to life how men behave in battle. He was often asked 'Why did they do it?'. He always replied 'Because they were Green Howards' and he meant it. He was then running a pub named 'The Green Howards'; his grandson was afterwards to serve in the Regiment.
17. Note 16, Hull.
18. Note 16, Hastings.
19. Synge and Clay are supplemented by *GHG* v 76, 1968, 145-6, 'Crossing of the Albert Canal – 7/8 September 1944' by Major-General R.K. Exham. At the time, Lieutenant-Colonel Exham of the Duke of Wellington's Regiment was in command, having taken over from Hastings when the latter was wounded in Normandy. Exham was later Colonel of the Dukes.
20. Howell, Tape 6.

Chapter 11. Retreat from Empire, and the National Serviceman: 1945-68 219-38

1. Blaxland, 506. About half of these losses occured in Korea; those wounded but not detained in hospital are not included.
2. Ibid., 18-19; *GHG*, v 56, 1946, 94, '2nd Battalion Notes'.
3. *GHG* v 58, 1950, 213-4, 'Location List of

Serving Officers'.

4. The main source of Malaya is Oldfield. Blaxland and O'Ballance are also useful. GhG, v 52, 1949, 169-71, has an article 'Green Howards in Anti-Terrorist Campaign (*sic*) in Malaya' by B. Russell-Jones (a District Officer).
5. Field Marshal Sir Nigel Bagnall, Tape 68.
6. Oldfield, xxvi.
7. Ibid., 45.
8. Ibid., 162.
9. Blaxland, 105-7
10. Oldfield, 155
11. Brigadier A. D. Miller, tape 49A; he commanded the 1st Battalion, 1952-5.
12. As well as Blaxland and Crawshaw, '2nd Battalion Notes' in the *GHG* and the author's personal knowledge and that of Lieutenant-Colonel E.D. Sleight are used for the Cyprus tour. *The Diehards: The Journal of the Middlesex Regiment,* v 12, 1956, 152-170, describes the arrival of the 1st Middlesex at Famagusta.
13. GHG, v 67, 1959, 'Review of the 1st Battalion in Hong Kong, 1956-1959'; Colonel H.A. Styles, Tape 67; he commanded the 1st Battalion, 1955-8.
14. GHG, v 70, 1962, 292-3, 'The 1st Battalion in BAOR 1959-62'.
15. Ibid., 279-80, 'The End of National Service' by J.B.O. (Brigadier J.B. Oldfield).
16. So called because his furniture is marked with a small mouse.
17. Information from the late Colonel J.M. Forbes, Regimental Secretary, 1961-78; GHG, v 79, 1971, 803.
18. GHG v 33, 1925. 117, 'Editorial'.

Chapter 12. Professionals Again: 1968-91 239-54

1. GHG, v 79, 1971, 943, 'Editiorial'.
2. Stanhope, Henry, *The Soldiers – An Anatomy of the British Army*, Hamish Hamilton, London, 1979, 36.
3. Ferrar, Bygone Days, 117.
4. Ibid., 122-3.

5. GHG, v82, 1974, 4, 'Editorial'.
6. Lieutenant-Colonel P.A. Inge, Tape 60A.
7. The author admits that the reference has eluded him.
8. *London Gazette* of 5 October 82.
9. Poll on Regimental Demography conducted by 1st Battalion from a sample of 436 soldiers, 7-11 January 1991.
10. GHG, v 90, 1983, 1, Reproduction of letter from HM the King.

Chapter 13. The New World Order: 1991-2001. 255-268

1. Excerpt from a US Information Team documentary film; "Operations in Bosnia" 1997.
2. FGHM Newsletter Issue 11, April 2001, page 5.
3. FGHM Newsletter Issue 11, April 2001, page 6.
4. FGHM Newsletter Issue 11, April 2001, page 9.
5. Field Marshal Lord Inge KG GCB DL – past Chief of the Defence Staff
   Major General FR Dannatt CBE MC – Commander NATO Allied Rapid Reaction Corps as Lt Gen in 2003.
   Brigadier JNR Houghton CBE – Chief of Staff to ARRC as Maj Gen in 2002.
   Brigadier CJ Marchant Smith CBE – late Commander 15 (North East) Brigade
   Brigadier JSW Powell OBE – late Commander 43 (Wessex) Brigade.
   Brigadier JCL King MBE – Defence Attaché British Embassy Seoul
   Brigadier AP Farquhar MBE – Commander 15 (North East) Brigade
   Brigadier DM Santa-Olalla DSO MC – Commander 2 Infantry Brigade

Epilogue 255-8

1. Henry Stanhope, *The Soldiers – An Anatomy of the British Army*, Hamish Hamilton, 1979, 36. Quoted from the *Daily Mail*, 17 December 1904.
2. *The Times*, 18 January 1991, 'British Infantry' by Philip Jacobson

Above: *Brigadier G. W. Eden, Colonel of the Regiment 1959-65.*

Above: *Brigadier J. B. Oldfield, Colonel of the Regiment 1975-82.*

Below: *Field Marshal Lord Inge, Colonel of the Regiment 1982-1994.*

Below: *Major General F. R. Dannatt, Colonel of the Regiment 1994.*

APPENDICES

A. COLONELS-IN-CHIEF AND COLONELS OF THE REGIMENT

Colonels-in-Chief

HM Queen Alexandra	1914-25
HM Haakon VII, King of Norway	1942-57
HM Olav V, King of Norway	1958-91
HM Harald V, King of Norway	1992-

Colonels of the Regiment

Colonel Francis Luttrell	1688
General the Right Honourable Thomas Erle	1691
Brigadier-General George Freke	1712
Brigadier-General Richard Sutton	1712
General the Honourable Charles Howard	1738
Lieutenant-General Lord George Beauclerk	1748
General David Graeme	1768
Field Marshal Sir Samuel Hulse	1797
General Sir Hew Dalrymple, Baronet	1810
General Sir Thompkins Hilgrove Turner	1811
General Sir Warren Marmaduke Peacocke	1843
Lieutenant-General Charles Turner	1849

Field Marshal Sir William Rowan	1854
General Sir Abraham Josias Cloete	1861
General Sir Robert Onesipherous Bright	1886
Lieutenant-General Edward Chippindall	1896
Major-General William Spencer Cooper	1902
Lieutenant-General Sir William Edmund Franklyn	1906
General Sir Edward Stanislaus Bulfin	1914
General Sir Harold Franklyn	1939
Major-General Alfred Eryk Robinson	1949
Brigadier George Wilfrid Eden	1959
Major-General Desmond Spencer Gordon	1965
Brigadier John Briton Oldfield	1975
Field Marshal Lord Inge	1982
Major General Francis Richard Dannatt	1994-

B. BATTLE HONOURS

Malplaquet; Belle Isle; Alma; Inkerman; Sevastopol; Tirah; Relief of Kimberley; Paardeberg; South Africa 1899–1902. **Great War:** Ypres 1914, 1915, 1917; Langermarck 1914, 1917; Gheluvelt; Neuve Chapelle; St Julien; Frenzenburg; Bellewaarde Aubers; Festubert 1915; Loos; Somme 1916–18; Albert 1916; Bazentin; Pozières; Flers-Courcelette; Morval; Thiepval; Le Transloy; Ancre Heights; Ancre 1916; Arras 1917, 1918; Scarpe 1917–18; Messines 1917, 1918; Pilckem; Menin Road; Polygon Wood; Broodseinde; Poelcappelle; Passchendaele; Cambrai 1917, 1918; St Quentin; Hindenburg Line; Canal du Nord; Beaurevoir; Selle; Valenciennes; Sambre; France and Flanders, 1914–18; Piave; Vittorio Veneto; Italy 1917, 1918; Sulva; Landing at Suvla; Scimitar Hill; Gallipoli 1915; Egypt 1916; Archangel 1918. Afghanistan 1919.

Second World War: Otta; Norway 1940; Defence of Arras; Dunkirk 1940; Normandy Landing; Tilly-sur-Seulles; St Pierre-la-Vielle; Gheel; Nederijn; North-West Europe 1940, 1944-5; Gazala; Defence of Alamein Line; El Alamein; Mareth; Akarit; North America 1942-3; Landing in Sicily; Lentini; Sicily 1943; Minturno; Anzio; Italy 1943-4; Arakan Beaches; Burma 1945.

C. Victoria and George Cross Holders

Private S. Evans, 1st Battalion, 13 April 1855, *London Gazette* entry 23 June 1857.

Private (later Corporal) John Lyons, 1st Battalion, 10 June 1855, *London Gazette* entry 24 February 1857.

Sergeant Alfred Atkinson, 1st Battalion, 18 February 1900, *London Gazette* entry 8 August 1902.

8191 Corporal William Anderson, 2nd Battalion, 13 March 1915, *London Gazette* entry 22 May 1915.

Major Stewart Walter Loudoun-Shand, 10th Battalion, 1 July 1916, *London Gazette* entry 9 September 1916.

Second Lieutenant Donald Simpson Bell, 9th Battalion, 5 July 1916, *London Gazette* entry 9 September 1916.

12067 Private William Short, 8th Battalion, 6 August, 1916 *London Gazette* entry 9 September 1916.

Captain Archie Cecil Thomas White, MC, 6th Battalion, 27 September and 1 October 1916, *London Gazette* entry 26 October 1916

Captain David Philip Hirsch, 4th Battalion, 23 April 1917, *London Gazette* entry 14 June 1917.

S. Evans

John Lyons

Alfred Atkinson

William Anderson

Stewart Walter Loudoun-Shand

Donald Simpson Bell

William Short

Archie Cecil Thomas White

David Philip Hirsch

242697 Private Tom Dresser, 7th Battalion, 12 May 1917, *London Gazette* entry 27 June 1917.

42537 Corporal William Clamp, 6th Battalion, 9 October 1917, *London Gazette* entry 18 December 1917.

Lieutenant-Colonel Oliver Cyril Spencer Watson, DSO, 28 March 1918, *London Gazette* entry 8 May 1918.

Second Lieutenant Ernest Frederick Beal, 13th Battalion, 21 and 22 March 1918, *London Gazette* entry 4 June 1918.

9545 Private Henry Tandey, DCM, MM, 28 September 1918, *London Gazette* entry 14 December 1918.

13820 Sergeant William McNally, MM, 8th Battalion, 27 October 1918, *London Gazette* entry 14 December 1918.

Lieutenant-Colonel Derek Anthony Seagrim, 7th Battalion, 20/21 March 1943, *London Gazette* entry 13 May 1943.

Company Sergeant-Major Stanley Elton Hollis, 6th Battalion, 6 June 1944, *London Gazette* entry 17 August 1944.

Lieutenant William Basil Weston, 3 March 1945, *London Gazette* entry 15 May 1945.

George Cross Holders

438271 Lance-Sergeant T. E. Alder, 14, November 1930.

2926329 Private T. McAvoy, 15 March 1939

4388265 Corporal T. Atkinson, 15 March 1939.

TOM DRESSER

WILLIAM CLAMP

OLIVER CYRIL SPENCER WATSON

ERNEST FREDERICK BEAL

HENRY TANDEY

WILLIAM McNALLY

DEREK ANTHONY SEAGRIM

STANLEY ELTON HOLLIS

WILLIAM BASIL WESTON

D. Regimental Family Tree

THE REGULAR ARMY

Depot / Regimental Depot

1873 — DEPOT
BRIGADE DEPOT NO. 4 RICHMOND
1879 — 19TH DIST REGIMENTAL DEPOT

REGIMENTAL DEPOT
1939 — INFANTRY TRAINING CENTRE NO. 5 ITC
No 19 PTC HQ REGIMENTAL DEPOT
1947 — REGIMENTAL DEPOT REACTIVATED
1951 — DISBANDED; THE GREEN HOWARDS REGIMENTAL HEADQUARTERS
1961 — FORMED

HOME GUARD
1939–45 NORTH RIDING HG
1ST NORTHALLERTON
2ND WHITBY
3RD GUISBOROUGH
4TH REDCAR
5TH MALTON
6TH YORK
7TH KIRBYMOORSIDE
8TH MIDDLESBROUGH
9TH
10TH SCARBOROUGH
11TH LEYBURN
12TH RICHMOND
13TH SALTBURN
STOOD DOWN 1945
1952 REFORMED
1ST NORTHALLERTON
2ND WHITBY
3RD REDCAR
4TH MALTON
5TH MIDDLESBROUGH
6TH SCARBOROUGH
7TH LEYBURN
8TH RICHMOND
9TH YARM
STOOD DOWN 1955

(1ST BATTALION) 1688 LUTTRELL'S REGIMENT
RAISED AT EXETER (DEVON) 19 NOVEMBER BY COLONEL FRANCIS LUTTRELL

1688 — ENGLAND (PORTSMOUTH, PLYMOUTH, ISLE OF WIGHT)
1691 — EARLE'S REGIMENT
1692 — FLANDERS, NAMUR
1696
1697 — ENGLAND
1697 — FLANDERS
1702 — IRELAND, ISLE OF WIGHT
1702 — CADIZ
1703 — GUADALOPE
1704 — IRELAND
1707 — ENGLAND
1707 — FLANDERS, MALPLAQUET
1714 — DOUAI, BETHUNE, BOUCHAIN
1712 — FREKE'S REGIMENT
1712 — SUTTON'S REGIMENT
1714 — ENGLAND
1718 — IRELAND
1715 — GROVE'S REGIMENT
1714 — VIGO EXPEDITION
1729 — SUTTON'S REGIMENT
1718 — IRELAND
1744 — ENGLAND
1738 — HOWARD'S REGIMENT
1744 — FLANDERS (FONTENOY)
1745 — ENGLAND
1746 — FLANDERS (ROUCOUX AND LAUFFELDT)
1749 — ENGLAND
1748 — BEAUCLERK'S REGIMENT
1749 — GIBRALTAR
1752
1751 — 19TH REGIMENT OF FOOT
1752 — ENGLAND
1761 — BELLE ISLE EXPEDITION
1763 — ENGLAND
1763 — GIBRALTAR
1771
1771 — ENGLAND AND IRELAND
1781 — AMERICAN WAR OF INDEPENDENCE
1782 — 19TH (1ST YORK NORTH RIDING REGIMENT)
1782 — JAMAICA
1791 — ENGLAND
1791
1793 — FLANDERS AND ENGLAND
1794 — FLANDERS
1795
1796 — CEYLON, TRAVANCORE (1809) AND CAPTURE OF KANDY (1815)
1820 — INDIA
1820 — ENGLAND
1826 — IRELAND
1826 — WEST INDIES
1836
1836 — IRELAND
1840
1840 — MALTA
1843 —

Regular Army column

1689 — 2ND BATTALION (RAISED DORSET 8 MARCH)
1689 — DUNDALK
1690 — BOYNE, WATERFORD
1691 — ATHLONE, AUGHRIM, GALLWAY
1692 — ENGLAND
1697 — DISBANDED 21 SEPTEMBER 1697
1756 — REFORMED AT MORPETH, 25 AUGUST
1758 — FORMED INTO 66TH FOOT (2ND BATTALION ROYAL BERKSHIRE REGIMENT (PRINCESS CHARLOTTE OF WALES'S OWN))
REFORMED AT EXETER (8 MARCH)
1858 — ENGLAND, IRELAND
1858
1863 — INDIA
1863
1876
1877 — ENGLAND
1881 — IRELAND
1881
1886 — ENGLAND
1886
1889
1890 — INDIA
1906 — TIRAH EXPEDITION
1909 — SOUTH AFRICA
1909 — ENGLAND
1913
1914 — FRANCE, FLANDERS (FIRST WORLD WAR)
1919 — IRELAND
1919
1925 — ENGLAND
1925 — WEST INDIES
1927 — EGYPT
1929
1929 — SHANGHAI
1930
1930 — POONA
1934
1934 — INDIA
1938 — WAZIRISTAN (37-38)
1939 — INDIA
1947 — EGYPT, SUDAN
1948
1949 — DISBANDED 31 MARCH; BECOMES 1ST BATTALION
1952 — RAISED AT BARNARD CASTLE (3 APRIL)
1953 — EGYPT
1954 — CYPRUS
1956 — ENGLAND (PLACED IN SUSPENDED ANIMATION)

VOLUNTEER FORCES

NORTH RIDING VOLUNTEER REGIMENTS OF NAPOLEONIC WARS 1796-1815

1858 LOCAL RIFLE VOLUNTEER CORPS OF THE NORTH RIDING
1860 THE NORTH RIDING RIFLE VOLUNTEER CORPS
(1ST ADMIN BATTALION, 2ND ADMIN BATTALION)
1880 BECOME THE NORTH YORK RIFLE VOLUNTEERS
(1ST BATTALION, 2ND BATTALION)
1883 BECOME VOLUNTEER BATTALIONS, OF THE REGIMENT
(1ST VOLUNTEER BATTALION, 2ND VOLUNTEER BATTALION)
1908 BECOME TERRITORIAL BATTALIONS OF THE REGIMENT

4TH BATTALION
1914 EMBODIED
1915 FRANCE AND FLANDERS
1918
1914 2ND/4TH BATTALION, 18TH (HS) BATTALION, 1917
1915 3RD/4TH BATTALION; AMALGAMATED WITH 3RD/5TH
BATTALION TO BECOME
1916 4TH (RESERVE) BATTALION
1939 EMBODIED
1940 FRANCE AND FLANDERS (RETREAT TO DUNKIRK) ENGLAND
1941 EGYPT, CYPRUS, WESTERN DESERT
1942 WESTERN DESERT (BATTALION CAPTURED; TO CADRE)
1947 REFORMED 1 MARCH
1961 BECOME 4TH/5TH BATTALION

5TH BATTALION
1914 EMBODIED
1915 FRANCE AND FLANDERS
1918
1914 2ND/5TH (ITS) BATTALION DISBANDED 1918
1915 3RD/5TH (TRAINING) BATTALION AMALGAMATED WITH 3RD/4TH
BATTALION (1916)
1939 EMBODIED
1940 FRANCE AND FLANDERS (RETREAT TO DUNKIRK) ENGLAND
1941 EGYPT, CYPRUS, WESTERN DESERT
1942 WESTERN DESERT (BATTALION CAPTURED; TO CADRE)
1949 5TH BATTALION COLOURS HANDED OVER
1951 631 LIGHT REGIMENT RA (gu) TA
631 (GH) LIGHT REGIMENT RA (TA)
1961 AMALGAMATED WITH 4TH BATTALION TO BECOME 4TH/5TH BATTALION;
COLOURS RETURNED
1967 4TH/5TH BATTALION DISBANDED 31 MARCH
1992 4TH/5TH BATTALION REFORMED
1993 4TH/5TH BATTALION DISBANDED. A AND B COMPANIES REMAIN

6TH BATTALION
1939 EMBODIED
1940 FRANCE AND FLANDERS (RETREAT TO DUNKIRK) ENGLAND
1941 CYPRUS, IRAQ, WESTERN DESERT, ALAMEN AKART
1943 SICILY, ENGLAND
1944 NORMANDY LANDING, BELGIUM, HOLLAND
1945 ENGLAND
1946 CYPRUS, DISBANDED
1947 BECOMES 87 MOVEMENT LIGHT BATTERY RA
1955 DISBANDED

7TH BATTALION (RAISED 1938, BRIDLINGTON)
1939 EMBODIED
1940 FRANCE AND FLANDERS (RETREAT TO DUNKIRK) ENGLAND
1941 CYPRUS, IRAQ, WESTERN DESERT, ALAMEN AKART
1943 SICILY, ENGLAND
1944 NORMANDY LANDING, BELGIUM, HOLLAND
1945 DISBANDED

1856 ENGLAND
1857 INDIA (MUTINY) (HAZARA, 1868)
1871 ENGLAND
1875 **19TH OR 1ST YORK NORTH RIDING (THE PRINCESS OF WALES'S OWN)**
1877 BERMUDA
1881 **THE PRINCESS OF WALES'S OWN (YORKSHIRE REGIMENT)**
1881 CANADA
1884 MALTA
1888 EGYPT (SUDAN EXPEDITION)
1888 CYPRUS
1889 ENGLAND
1892 JERSEY
1895 IRELAND
1899 SOUTH AFRICA (WAR)
1902 **ALEXANDRA, PRINCESS OF WALES'S OWN YORKSHIRE REGIMENT**
1902 ENGLAND
1908 EGYPT
1912 INDIA (THIRD AFGHAN WAR)
1919 PALESTINE
1919 INDIA
1920 **THE GREEN HOWARDS (ALEXANDRA, PRINCESS OF WALES'S OWN YORKSHIRE REGIMENT)**
1926 EGYPT
1927 ENGLAND
1927 HONG KONG
1928 SHANGHAI
1928 ENGLAND
1937 MALTA
1938 PALESTINE
1939 ENGLAND, FRANCE, NORWAY, NORTHERN IRELAND, INDIA, PERSIA
1939 PALESTINE, SICILY, ITALY, EGYPT, GERMANY
1945 WEST GERMANY
1947 ENGLAND
1949 SUDAN
1952 MALAYA
1953 AUSTRIA AND GERMANY
1956 HONG KONG
1959 WEST GERMANY
1963 LIBYA
1966 COLCHESTER
1969 WEST GERMANY
1974 BERLIN
1976 CHESTER
1978 NORTHERN IRELAND
1980 CATTERICK
1983 WEST GERMANY
1987 NORTHERN IRELAND
1989 CATTERICK
1994 GERMANY
2000 WARMINSTER

MILITIA
1759 RICHMONDSHIRE BATTALION, CLEVELAND AND BULMER BATTALION
NORTHUMBERLAND
1761 HEXHAM RIOTS
1778 BATTALION AMALGAMATED; EMBODIED AT LEEDS
1779 YORK, KENT
1780 GOSPORT, LONDON
1781 HM 13TH OR NORTH YORKSHIRE REGIMENT OF MILITIA
1783 DISEMBODIED AT RICHMOND
1793 EMBODIED AT RICHMOND
1802 DISEMBODIED 23 APRIL
1803 EMBODIED AT RICHMOND; SUNDERLAND
1805 WEYMOUTH
1808 SOUTH COAST
1813 IRELAND
1814 NORTH YORK LIGHT INFANTRY REGIMENT OF MILITIA
1816 DISEMBODIED
1852 NORTH YORKSHIRE RIFLES (MILITIA)
1854 EMBODIED AT RICHMOND
1856 BRADFORD; DISEMBODIED AT RICHMOND 17 JUNE
1881 BECOME MILITIA BATTALIONS OF THE REGIMENT

3RD MILITIA BATTALION (5 W. YORKS)
1899 EMBODIED AT RICHMOND
1900 SOUTH AFRICA
1902 RETURNED TO UK
1908 BECOME SPECIAL RESERVE BATTALIONS
1914 ENGLAND
1918
1919 IRELAND, INTO SUSPENDED ANIMATION; EXISTED IN NAME
ONLY; DISBANDED 1953

4TH MILITIA BATTALION (NORTH YORK) (CEASED TO BE A RIFLE REGIMENT)
1899 EMBODIED AT STRENSALL, SHEFFIELD
1901 DISEMBODIED AT SHEFFIELD
1902 EMBODIED AT RICHMOND (FEBRUARY), SOUTH AFRICA
1908 DISBANDED

1914-1918 WAR BATTALIONS
1S GARRISON BATTALION, INDIA (1915-19)
2ND (HS) GARRISON BATTALION
6TH (S) BATTALION, GALLIPOLI, EGYPT, FRANCE, FLANDERS, RUSSIA
7TH (S) BATTALION, FRANCE, FLANDERS
8TH (S) BATTALION, FRANCE, FLANDERS, ITALY
9TH (S) BATTALION, FRANCE, FLANDERS, ITALY
10TH (S) BATTALION, FRANCE, FLANDERS
11TH (HS) BATTALION, AMALGAMATED WITH 18TH BATTALION 1916
12TH (S) BATTALION, (PIONEERS), FRANCE, FLANDERS
13TH (S) BATTALION, FRANCE, FLANDERS, RUSSIA
14TH (RES) BATTALION, AMALGAMATED WITH 81 TRAINING RESERVE BATALION 1916
15TH (RES) BATTALION, YOUNG SOLDIERS DISBANDED 1916
16TH AND 17TH LABOUR BATTALIONS MERGED INTO LABOUR CORPS 1917
18TH (HS) TERRITORIAL BATTALION

1939-45 WAR BATTALIONS
1ST/8TH (HD) BATTALION
2ND/8TH (HD) BATTALION BECOMES 13TH (HD) BATTALION
8TH/13TH (HD) BATTALION AMALGAMATED TO FORM 30TH BATTALION
30TH BATTALION UK, ALGIERS, ITALY
9TH BATTALION BECOMES 108 LAA REGIMENT RA (1941)
10TH BATTALION BECOMES 12TH (YORKSHIRE) PARACHUTE REGIMENT (1943)
11TH BATTALION (TRAINING) DISBANDED 1945
12TH BATTALION FORMED AS 5 HOLDING BATTALION, BECOMES 50TH BATTALION,
CONVERTED TO 161 RECCE REGIMENT 1943.

E. The Green Howards' Collect

Lord Jesus Christ, Shepherd of our Souls, so lead, we pray thee, The Green Howards, that as we bear the symbol of Thy Cross we may be always ready to take up our Cross and follow thee, till, having passed through the valley of the shadow of death we may rest in Green Pastures in Thy care, who art with the Father and the Holy Ghost one God, world without end. AMEN.

F. Regimental Music

The Quick March: *The Green Howards*
(*The Bonnie English Rose*) Arr. C. D. Jarrett

1st B♭ Cornet

The Grand or Slow March

The Funeral March

G. Regimental Secretaries

1945–60	Major H. W. Ibbetson
1960–78	Colonel J. M. Forbes, DL, JP
1978–81	Lieutenant-Colonel G. T. M. Scrope, OBE, DL
1981–2	Lieutenant-Colonel D. M. Stow, MBE
1982–6	Lieutenant-Colonel D. J. Bottomley, MBE
1986–	Lieutenant-Colonel N. D. McIntosh, MBE

List of Abbreviations

ADC	Aide-de-Camp
AQ	Army Quartely
ARRC	Allied Rapid Reaction Corps
BEF	British Expeditionary Force
CLR	Ceylon Literary Register
CQMS	Company Quartermaster Sergeant
CSM	Company Sergeant-Major
DCM	Distinguished Conduct Medal
DSO	Distinguished Service Order
DNB	Dictionary of National Biography
FGHM	Friends of the Green Howards Museum
GHG	*Green Howards' Gazette*
GHQ	General Headquarters
GOC	General Officer Commanding
GOC-in-C	General Officer Commanding-in-Chief
HMSO	Her Majesty's Stationery Office
IFOR	NATO Implementation Force
IRA	Irish Republican Army
IWM	Imperial War Museum
JRASCB	Journal of the Royal Asiatic Society (Ceylon Branch)
KFOR	Kosovo Force
MM	Military Medal
MC	Military Cross
NAM	National Army Museum
NATO	North Atlantic Treaty Organisation
NCO	Non-Commissioned Officer
OTC	Officers' Training Corps
PRO	Public Record Office
QM	Quartermaster
RA	Royal Artillery
RE	Royal Engineers
RIC	Royal Irish Constabulary
RHQ	Regimental Headquarters
RQMS	Regimental Quartermaster Sergeant
RUC	Royal Ulster Constabulary
RSM	Regimental Sergeant-Major
RUSI	Royal United Services Institute for Defence Studies Journal
SEP	Surrended Enemy Personnel
TA	Territorial Army
UN	United Nations
USI	United Services Institute
VC	Victoria Cross
WO	War Office or Warrant Officer

BIBLIOGRAPHY

PRINTED WORKS

Regimental (This does not include books listed under chapters. Those written by Green Howards are marked*. Mention of Ferrar alone in the Notes signifies his *Historical Record . . .*)

*Calladine, Colour-Sergeant George. *Diary*, Eden Fisher, 1922

Cannon, Richard. *Historical Record of the 19th or the First Yorkshire North Riding Regiment of Foot*, Parker, Furnival & Parker, 1848

*Ferrar, Major M. L., *A History of the Services of the 19th Regiment from its Formation In 1688 To 1911*, Eden Fisher, 1911

*— Bygone Days, R. Carswell, Belfast, 1924.

*— Officers of The Green Howards, 1688 to 1931, W. & G. Baird, Belfast, 1931.

*— Selections from the *Green Howards' Gazette*, Baird, Belfast, 1934.

*Fife, Lieutenant-Colonel Ronald. *Mosaic of Memories*, Heath Cranton, 1943

*Forbes, Colonel J. M. *Addendum to Officers of the Green Howards*, privately printed, 1971.

The Green Howards' Gazette, vols 1–99

*Powell, Geoffrey. *The Green Howards*, Hamish Hamilton, 1968

*Turton, Robert Bell. *The History of the North York Militia*, Patrick & Shotton, 1973

Wylly, Colonel H. C. *The Green Howards in the Great War*, Privately printed, 1926

General

Cole, Major D. H., and Priestley, Major E. C. *An Outline of British Military History 1660–1937*, 1937.

Barnett, Correlli. *Britain and Her Army, 1509–1970*, 1970

Dennis, Peter. *The Territorial Army, 1906–1940*, Boydell Press, 1987

Dictionary of National Biography: Compact Edition, OUP, 1975

Norman, C. B. *Battle Honours of the British Army*, John Murray, 1911

Spiers, Edward M. *The Army and Society, 1815–1914*, Longman, 1980

Trevelyan, G. M. *History of England*, Longman, 1926

Chapters 1 and 2

Atkinson, C. T. *Marlborough and the Rise of the British Army*, Putnam, 1921

Boyer, A. *The History of the Reign of Queen Anne – Year the Eighth*, T. Ward, 1710

Chandler, David. *The Art of Warfare in the Age of Marlborough*, Batsford, 1976

—(ed.) *The Marlborough Wars: Robert Parker and Comte De Merode-Westerloo*, Longmans, 1968

Churchill, Winston S. *Marlborough: His Life and Times*, vols. 1–4, new ed., 1934

Dalton, Charles. *English Army Lists and Commission Registers*, vol. 3, 1896

D'Auvergne, Edward. *History of the Campaign in the Spanish Netherlands – 1693*, 1693

— *History of the Campaign in the Spanish Netherlands – 1694*, 1694

— *History of the Last Campaign in the Spanish Netherlands – 1695*, 1695

Dixon, Frederick. 'Landen', in *Temple*

Bar Magazine, vol. 117, 1899, p.174

Fortescue, The Hon J. W. *A History of the British Army*, vols. 1 – 3, 1899–1902

Guy, Alan J. and Spencer-Smith, Jenny (eds). *Glorious Revolution: The Fall and Rise of The British Army, 1660–1704*, National Army Museum, 1988

Hargreaves, Reginald. *The Bloodybacks: The British Serviceman in North America and The Caribbean, 1655–1783*, 1968

Holbroke, Lieutenant Richard. 'The Siege and Capture of Belle-Isle – 1761', *USJ*, vol. 43, Feb 1899

Kane, Richard. *Campaigns of King William and Queen Anne – 1689–1712*, 1745

Little, Bryan. *The Monmouth Episode*, Werner Laurie, 1956

Lyte, Sir H. C. Maxwell. *A History of Dunster*, Part 1, St. Catherine Press, 1909

McCrady, Edward. *The History of South Carolina in the Revolution*, Macmillan, 1902

Rogers, Colonel H. C. B. *The British Army of the Eighteenth Century*, 1977

Scouller, Major R. E. *The Armies of Queen Anne*, Clarendon Press, 1966

Skrine, Francis Henry. *Fontenoy and the War of the Austrian Succession*, Blackwood, 1906

Story, George. *An Impartial History of the Wars in Ireland*, Parts 1 & 2, 1693

Ward, Christopher. *The War of the Revolution*, Macmillan, New York, 1952

Chapter 3

*Anderson, Captain Thomas Ajax. *Poems Written Chiefly in India*, Philanthropic Society, 1809

*— *A Wanderer in Ceylon*, T. Egerton Military Library, 1817

Letters of Captain Herbert Beaver (HM 19th Regt) March-April 1803, CLR, 4 (2), Oct 1918, 6–31

Sketches of the Services of the Late Major Herbert Beaver, *USJ*, Part II, pp.431–8; 705–15, 1829

Cordiner, Revd. James. *A Description of Ceylon*, two vols., Longman, Hurst, Rees, Orme & Brown, 1807

Davy, John. *An Account of the Interior of Ceylon*, Longman, Hurst, Rees, Orme

& Brown, 1821

Fortescue, The Hon J. W. *A History of the British Army*, vols. 3–5, 1902–10

Howell, John (ed.). *The Life of Alexander*, vol. 1, T. Egerton Military Library, 1830

*Johnston, Major Arthur. *Narrative of an Expedition to Candy in the Year 1810*, Wm. S. Orr, 1810

Marshal, Dr Henry. *Notes on the Medical Topography of the Interior of Ceylon*, Burgess & Hill, 1821

*Percival, Captain Robert. *An Account of the Island of Ceylon*, G. & R. Baldwin, 1805

*Powell, Geoffrey. *The Kandyan Wars: The British Army in Ceylon, 1803–18*, Leo Cooper, 1973

Reith, Charles. *An Ensign of the 19th Foot*, Heath Cranton, 1925 (Fiction).

— *The 19th (Yorkshire) Regiment and its Connection with Ceylon*, JRASCB, 23(66), 1913, pp.55–69

*Turton, Robert Bell. *The History of the North York Militia*, Patrick & Shooton, 1973

Western, I. R. *The English Militia in the 18th Century*, Routledge & Kegan Paul, 1965

Chapter 4

*Ferrar, Major M. L. *Selections From GHG* (in the Crimea with the 19th Foot by Margaret Kerwin).

Fortescue, The Hon J. W. *A History of the British Army*, vol. 13, Macmillan, 1930

Hibbert, Christopher. *The Destruction of Lord Raglan*, Longmans, 1961

Kinglake, Alexander William. *Invasion of the Crimea*, vols. 1–8, Blackwood, 1863–87

Russell, William Howard. *Russell's Despatches from the Crimea*, ed. Nicholas Bentley, Panther edition, 1970

*Usherwood, C. W. *Service Journal of Charles William Usherwood, 1852–1864*, privately printed, no date

Chapter 5

Amery, L. S. (ed.). *The Times History of the War in South Africa 1899–1902*, vol.

3, Sampson Low & Marston, 1905

Calwell, Colonel C. E. *Tirah 1897*, Constable, 1911.

Colville, Colonel H. E. *History of the Sudan Campaign*, Part 2, Eyre & Spottiswoode, 1889

Farwell, Bryon. *The Great Boer War*, Allen Lane, 1977

*Ferrar, Major M. L. *With The Green Howards in South Africa 1899–1902*, Eden Fisher, 1904

Harries-Jenkins, Gwyn. *The Army in Victorian Society*, Routledge & Kegan Paul, 1977

Maurice, General Sir Frederick. *History of the War in South Africa*, vol. 2, Hurst & Blackett, 1907

*Mott, Edward Spencer ('Nathaniel Gubbins'). *A Mingled Yarn*, Edward Arnold, 1898

Nevill, Captain H. L. *Campaigns on the North-West Frontier*, John Murray, 1912

Royle, Charles. *The Egyptian Campaigns, 1882 to 1885* (new and rev. ed.), Hurst & Blackett, 1900

Frontier & Overseas Expeditions From India, vol. 1 Tribes North of Kabul, Simla, 1907. Compiled by the Intelligence Division, Army Headquarters, India

Chapters 6 and 7

Anon. *History of the 50th Infantry Brigade 1914–1919*, privately printed, 1919

Aspinall-Oglander, Brigadier-General C. F. *Military Operations, Gallipoli*, Heinemann, 1932

Atkinson, C. T. *The Seventh Division, 1914–1918*, John Murray, 1927

Atteridge, A Hilliard. *History of the 17th (Northern Division)*, Glasgow University Press, 1919

Edmonds, Brigadier-General J. E. et al., *Military Operations, France and Belgium, 1914–15*, vols. 2, 3, 4 1916; vols. 1, 2 1917; vols 1, 2, 3 1918; vols. 3, 4, 5, 6; Macmillan and HMSO, 1917–48

— *Military Operations, Italy*, HMSO, 1949

Liddell Hart, B. *A History of the World War, 1914–1918*, Faber, 1934

Moorehead, Alan. *Gallipoli*, Hamish Hamilton, 1956

*Read, Sir Herbert. *The Contrary Experience*: Autobiographies, Faber, 1963

Robertson, William (ed.). *Middlesbrough's Effort in the Great War*, Jordison, Middlesbrough, no date

Simkins, Peter. *Kitchener's Army: The Raising of the New Armies, 1914–18*, Manchester University Press, 1988

Spiers, Edward M. *The Army and Society, 1815–1914*, Longman, 1980

Terraine, John. *The First World War, 1914–1918*, Hutchinson, 1965

— *The Road to Passchendaele: The Flanders Offensive of 1917*, A Study in Inevitability, Leo Cooper, 1977

— *To Win a War 1918: The Year of Victory*, Sidgwick & Jackson, 1978

Wyrall, Everard. *The History of the Fiftieth Division, 1914–1919*. Lund, Humphries & Co., 1939

Wylly, Colonel H. C. *The Green Howards in the Great War*, Privately printed, 1926

Chapter 8

Bond, Brian. *British Military Policy Between the Two World Wars*, Clarendon Press, 1980

Halliday, E. M. *The Ignorant Armies: The Anglo-American Archangel Expedition*, Weidenfeld & Nicholson, 1961

Holt, Edgar. *Protest in Arms: The Irish Troubles, 1916–1923*, Putnam, 1960

Ironside, Edmund. *Archangel 1918–1919*, Constable, 1953

Lowe, Major T. A. *Some Reflections of a Junior Commander upon the 'Campaign' in Ireland 1920 & 1921*, AQ, vol. 5, 1922, pp.50–8.

Marlowe, John. *The Seat of Pilate: An Account of the Palestine Mandate*, Cresset Press, 1959

Molesworth, Lieutenant-General G. N. *Afghanistan 1919*, Asia Publishing House, 1962

Townshend, Charles. *The British Campaign in Ireland, 1919–1921*, Oxford University Press, 1975

Swinson, Arthur. *North-West Frontier*, Hutchinson, 1967

Wylly (*see* Chapter 7 above)

Chapters 9 and 10

Aris, George. *The Fifth British Division, 1939 to 1945*, privately printed, 1959

*Clay, Ewart W. *The Path of the 50th: The Story of the 50th (Northumbrian) Division in the Second World War*, Gale & Polden, 1950

Bush, Lieutenant-Colonel W. E. *150th Infantry Brigade in the Middle East: June 1941–June 1942*, privately printed, 1944

*Franklyn, General Sir Harold. *The Story of One Green Howard in the Dunkirk Campaign*, privately printed, 1966

Fraser, David. *And We Shall Shock Them: The British Army in the Second World War*, Hodder & Stoughton, 1983

*Hull, C. Macdonald. *A Man From Alamein*, Corgi, 1973

*Mander, D'A. *Mander's March on Rome*, Alan Sutton, 1987

Parker, R. A. C. *Struggle For Survival: The History of the Second World War*, OUP, 1989

Synge, Captain W. A. T. *The Story of the Green Howards, 1939–1945*, Privately printed, 1952

*Trevelyan, Raleigh. *The Fortress: A Diary of Anzio and Afterwards*, Collins, 1956 (he was Rifle Brigade seconded to the Green Howards)

Chapter 11

Blaxland, Gregory. *The Regiments Depart: A History of the British Army 1945–1970*, Kimber, 1971

Crawshaw, Nancy. *The Cyprus Revolt: An Account of the Struggle for Union with Greece*, Allen & Unwin, 1978

O'Ballance, Edgar. *Malaya: the Communist Insurgent War, 1948–60*, Faber & Faber, 1966

*Oldfield, Major J. B. *The Green Howards in Malaya (1949–1952)*, Gale & Polden, 1953

Official Papers

British Minor Expeditions 1746 to 1814, HMSO, 1884

Green Howards Claim for Battle Honours submitted in May 1935 in accordance with War Office Letter 20/Gen/5408 (AG4d) dated 17 February 1934

Historical Record of 2nd Bn The Green Howards 1858–1942

London Gazette No 2892 of 27–31 July 1693, 'The Official Account of the Battle of Landen'

London Gazette, April and May 1761. Extraordinary October 1746

Other Material

*Erle Papers (Churchill College, Oxford) [Transcription by Lieutenant-Colonel J. R. Neighbour, RHQ The Green Howards]

*Letter Books of Major Macdonald, 1815–20. (RHQ, The Green Howards. Part printed in GHG vol. 9, 1901, pp.8–9; 24–7)

Green Howard Archives

Green Howard Sound History Archives

Green Howards Museum Archives

Nugent Papers, NAM

Peasley, Una, *Edmund Lockyer and the Second British Empire*, Master of Letters Thesis for University of New England, 1990

White, Colonel A. C. T. Correspondence with author

PRO, WO 167

INDEX

With the exception of some well-known senior officers, officers, NCOs and other ranks listed are Green Howards unless otherwise indicated. Ranks given are the highest known to have been reached. Page numbers in italics refer to illustrations.

236; Princess Alexandra and Danish/Norwegian connection, 101, 110, 189, 217-18, 232-3, 254; Canadian units, links, 15, 217-18; Freedoms of Boroughs, *160*, 227, 254; Guards of Honour, *175*, *230*, *233*, 234, *249*, 249, 256; Officers' Dinner, 110, *174*, 237, 256, *271*; Richmond Sunday, *217*, 236; tercentenary, 13, 79-80, *225*, 252-3, 270; Tower of London, 259; *see also* Yorkshire, Regt's ties with

awards for bravery: pre-WW1, 92, 114, 129; 1st WW, 109, 138, 139-40, 150, 151, 154, 157, 159, *161*, 161, 162, 163; inter-war, 178, 182; 2nd WW, 191, *193*, 198, 205, 209, 210, *212*, 215, 216; Ireland, 242, 243, 244, 248, 252, 269; Falklands, 254

Association and benevolent fund, 110, 236

Army Cadet Force, 236

band and march, 83-4, 101, *128-9*, 239, *246*, 247, 254, 255, 256

Regimental Council, 237

regimental journal and histories, 11, 13, 14, 106, 157, 223, 236

trophies and Museum, *32*, 90, *96*, 106, *114*, 126, *224*, 236, *243*, 243-4, *244*, 260

sport, 71, 82, 97, 102, 175, 202, 228, 257

Green Howards, history:

as Luttrell's Regiment (1688-90), 17-19, 20, 25, *32*

as Erle's Regiment (1689-1712),1st Bn (1690-7); raising, 20-1, 25-6; Continental wars and Ireland, 27, 29-31, 39-42; Cadiz, W. Indies, Newfoundland and home, 36-9

as Erle's Regiment, 2nd Bn (1690-7) Ireland and Continental wars, 21-2, 23, 24-5, 27, 34; disbandment, 34

as Freke's Regiment (1712), 43

as Sutton's Regiment (1712-15) home garrison and Ireland, 43-4

as Grove's Regiment (1715-29), Vigo expedition 43, 44

as Sutton's Regiment (1729-38), 43, 45

as Howard's Regiment (1738-48), Ireland and Flanders, 45, 46-8

as Beauclerk's Regiment (1748-58), designated 19th Foot and 2nd Bn formed, 48, 49

Belle Isle, (1761), 50-1, *51*

North America and Jamaica (1781-2), 53-6

Netherlands expedition (1793-4), 57-8

Ceylon and Kandyan wars (1796-1820), 59-71, 74-8

India (1799-1807), 60-1, 62-3, 70